ADVANCES IN HEAT EXCHANGERS

Edited by **Laura Castro Gómez**
and **Víctor Manuel Velázquez Flores**

Advances in Heat Exchangers
http://dx.doi.org/10.5772/intechopen.74640
Edited by Laura Castro Gómez and Víctor Manuel Velázquez Flores

Contributors

Imen Jmal, Mounir Baccar, Martín Picón-Núñez, Jorge Luis García-Castillo, Jorge Carlos Melo-González, Jaafar Albadr, Uwe Scheithauer, Kevin Noack, Martin F. Eichenauer, Daniel Lordick, Johannes Abel, Gregor Ganzer, Richard Kordaß, Mathias Hartmann, Laura Castro Gómez

Notice

Statements and opinions expressed in the chapters are these of the individual contributors and not necessarily those of the editors or publisher. No responsibility is accepted for the accuracy of information contained in the published chapters. The publisher assumes no responsibility for any damage or injury to persons or property arising out of the use of any materials, instructions, methods or ideas contained in the book.

First published in London, United Kingdom, 2019 by IntechOpen
IntechOpen is the global imprint of INTECHOPEN LIMITED, registered in England and Wales, registration number: 11086078, The Shard, 25th floor, 32 London Bridge Street
London, SE19SG – United Kingdom
Printed in Croatia

British Library Cataloguing-in-Publication Data
A catalogue record for this book is available from the British Library

Additional hard copies can be obtained from orders@intechopen.com

Advances in Heat Exchangers, Edited by Laura Castro Gómez and Víctor Manuel Velázquez Flores
p. cm.
Print ISBN 978-1-78985-073-4
Online ISBN 978-1-78985-074-1

We are IntechOpen,
the world's leading publisher of
Open Access books
Built by scientists, for scientists

4,000+
Open access books available

116,000+
International authors and editors

120M+
Downloads

Our authors are among the

151
Countries delivered to

Top 1%
most cited scientists

12.2%
Contributors from top 500 universities

Interested in publishing with us?
Contact book.department@intechopen.com

Numbers displayed above are based on latest data collected.
For more information visit www.intechopen.com

Meet the editors

Laura Castro Gómez has a PhD degree in Engineering and Applied Sciences. She is Professor-Researcher in Mechanical Engineering at Morelos State University, Mexico. Professor Castro teaches the subjects of thermodynamics, fluid mechanics, and heat transfer, among others. She has expertise in turbomachinery, fluid flows, and heat exchangers. She has published several papers in scientific journals such as *Engineering Failure Analysis*, *Energy Procedia*, *Advances in Mechanical Engineering*, etc. She has also published four book chapters and coedited one book. In addition, she is a reviewer of journals such as *International Journal of Energy Research* and *Heat and Mass Transfer*. She is an honorary member of the Mexican Society of Mechanical Engineering, AC (SOMIM).

Víctor Velázquez Flores is Professor of Chemical Engineering at Morelos State University, Mexico. He received the BChE degree and PhD in Applied Chemical Engineering from Morelos State University. Professor Velázquez has taught the subjects of thermodynamics, applied thermodynamics, applied fluid mechanics, mass and energy balance, and heat transfer. Professor Velázquez has published articles in refereed and indexed journals and has given workshops for the training of students and teachers. In addition, he has been director of several theses and served as a dissertation advisor. He is a member of the Mexican Institute of Chemical Engineers.

Contents

Preface IX

Section 1   **Introduction   1**

Chapter 1   **Introductory Chapter: Heat Exchangers   3**
Laura L. Castro Gómez

Section 2   **Design   7**

Chapter 2   **Numerical Investigation of PCM Melting in a Finned Tube Thermal Storage   9**
Imen Jmal and Mounir Baccar

Chapter 3   **Use of Heat Transfer Enhancement Techniques in the Design of Heat Exchangers   23**
Martín Picón-Núñez, Jorge C. Melo-González and Jorge Luis García-Castillo

Chapter 4   **Potentials and Challenges of Additive Manufacturing Technologies for Heat Exchanger   41**
Uwe Scheithauer, Richard Kordaß, Kevin Noack, Martin F. Eichenauer, Mathias Hartmann, Johannes Abel, Gregor Ganzer and Daniel Lordick

Section 3   **Fluids   63**

Chapter 5   **Thermal Performance of Shell and Tube Heat Exchanger Using PG/Water and Al2O3 Nanofluid   65**
Jaafar Albadr

Preface

Heat exchangers are devices extremely relevant for engineering, research, and industry in general. They have been used in several processes to increase or decrease the temperature of different working fluids. Because of this, any improvement developed for the aforementioned equipment helps to optimize the whole process at laboratory and industrial levels. There are many opportunity areas to modify heat exchangers, for example in their design, materials, even in the fluids of exchange.

In this sense, the present book, along with introductory chapter, compiles some advances in these issues in the matter of design (three chapters) and working fluids (one chapter). The first chapter presents advances in a finned heat exchanger for thermal storage. The second chapter presents investigations relative to techniques of heat transfer for heat exchangers. In the third chapter, advances for integrated structures in heat exchangers are presented. Finally, in fourth chapter, advances in the use of nanofluids for heat exchangers are presented.

Laura Castro Gómez and Víctor Manuel Velázquez Flores
Universidad Autonoma del Estado de Morelos (UAEM)
Cuernavaca, Mexico

Introduction

Chapter 1

Introductory Chapter: Heat Exchangers

Laura L. Castro Gómez

Additional information is available at the end of the chapter

http://dx.doi.org/10.5772/intechopen.83376

1. Introduction

Nowadays, engineering has a lot of challenges to improve or completely change equipment and devices used in industry, laboratory, and daily life. Examples of these devices are heat exchangers, which, as their name refer, exchange energy from a hot fluid to a cold fluid or vice versa. These devices could be as small as the cooling system of a CPU processor or as big as used in the industry of many meters of height and length.

Heat exchangers are used in almost all systems that could need heating or cooling of equipment or process fluids in industry.

2. Types of heat exchangers

Heat exchangers could be constructed in many arrangements, depending on their application. The most usual classification has been given by several authors like Shah [1] and Cengel [2]. This classification could be:

1. Contact of fluids

    a. Direct: could be of immiscible fluids, vapor-liquid or gas-liquid

    b. Indirect: could be of storage, fluidized bed

2. Flow of working fluids

    a. Single pass: counterflow, cross flow, parallel flow, etc.

    b. Multipass: extended surface, shell and tubes, and plates

3. Compactness

   a. Compact

   b. Noncompact

4. Construction

   a. Tubular: double pipe, shell and tube, and pipe coils

   b. Plate: spiral and plate coil

   c. Extended surface: finned tube or plates

   d. Regenerative: rotatory and mixing matrix

From this classification, it can be noted that there are several design options to select the exchanger that suits the needs of a process or the available space capacities.

Usually, this equipment is manufactured on materials with high heat transfer capacity, such as metals, but they can be designed for special operating conditions where metals have no good performance, like heavy fouling, highly viscous fluids, erosion, corrosion, and so on. So, they could be constructed with a variety of nonmetal materials: graphite, glass, and Teflon [3].

In the matter of working fluids, the same case of materials are used with good heat transfer capacities to maintain either its low or high temperature. Some fluids could be selected as coolants, such as brine; ammonia solutions; R-134; other organic fluids such as toluene, R-12, and Therminol; water; or others for heating systems [4, 5].

3. Advances on heat exchangers

As can be seen in the previous section, there are many opportunities to improve heat exchangers.

The present book compiles some advances in these issues. In the matter of design, three chapters and one chapter for working fluids are presented. In the first chapter, advances in a finned heat exchanger for thermal storage are presented. In in the second chapter, investigations relative to techniques of heat transfer for heat exchangers are presented. In the third chapter, advances for integrated structures in heat exchangers are presented. Finally, in the fourth chapter, advances in the use of nanofluids for heat exchangers are presented.

In the first chapter in design section, "Numerical Investigation of PCM Melting in a Finned tube Thermal Storage," authors propose a numerical investigation based on an enthalpy formulation to study the melting of a PCM in a finned heat exchanger. This numerical approach simultaneously gives the temperature distributions in the PCM storage system and temporal propagation of the solidification front during the solidification of the PCM when it is exposed to a cold air flow. Also, the transient evolution of the longitudinal air temperature profiles is given in this study.

The second chapter "Heat Transfer Enhancement in Tubular Heat Exchangers," by Martín Picón-Núñez et al. describes the concept of heat transfer enhancement and the ways it is applied to the development of new heat exchanger technology. Heat transfer enhancement refers to the application of basic concepts of heat transfer processes to improve the rate of heat removal or deposition on a surface. In the flow of a clean fluid through the tube of a heat exchanger, the boundary layer theorem establishes that a laminar sublayer exists where the fluid velocity is minimal. Heat transfer through this stagnant layer is mainly dominated by thermal conduction becoming the major resistance to heat transfer. From an engineering point of view, heat transfer can be enhanced if this stagnant layer is partially removed or eliminated. In single-phase heat transfer processes, three options are available to increase the heat transfer rate. One of them is the choice of smaller free-flow sectional area for increased fluid velocity bringing about a reduction of the thickness of the laminar sublayer. A second option is the engineering of new surfaces which cause increased local turbulence, and the third option consists in the use of mechanical inserts that promote local turbulence. The application of these alternatives is limited by the pressure drop.

In "Integrated Structures for Heat Exchangers" by Uwe Scheithauer et al., authors refer to the advantage of additive manufacturing (AM) technologies, enabling a radical paradigm shift in the construction of heat exchangers. In place of a layout limited to the use of planar or tubular starting materials, heat exchangers can now be optimized, reflecting their function and application in a particular environment. The complexity of form is no longer a restriction but a quality. Instead of brazing elements, resulting in rather inflexible standard components prone to leakages, with AM, we finally can create seamlessly integrated and custom solutions from monolithic material. To address AM for heat exchangers, we both focus on the processes, materials, and connections as well as on the construction abilities within certain modeling and simulation tools. AM is not the total loss of restrictions. Depending on the processes used, delicate constraints have to be considered. On the other hand, the materials used to manufacture heat exchangers with this technique could operate in a wide temperature range. It is evident that conventional modeling technics cannot match the requirements of a flexible and adaptive form finding. Instead, we exploit biomimetic and mathematical approaches with parametric modeling. This results in unseen configurations and pushes the limits of how we should think about heat exchangers today.

The section of working fluids, "Heat flow inside heat exchanger using Al_2O_3 nanofluid with different concentrations" by Jaafar Albadr, shows an experimental investigation on a forced convection heat flow and characteristics of a nanofluid containing water with different volume concentrations of Al_2O_3 nanofluid (0.3–2%) flowing inside a horizontal shell and tube heat exchanger in a counterflow under turbulent conditions. The Al_2O_3 nanoparticles of about 30 nm diameter are utilized. The results indicate that the convective heat transfer coefficient of nanofluid is higher than that of the base liquid at same inlet temperature and mass flow rate. The heat transfer coefficient of the nanofluid increases with the increase in mass flow rate. Furthermore, the heat transfer coefficient increases with the increase in the Al_2O_3 nanofluid volume concentration. Results illustrate that the increase in volume concentration of the nanoparticles leads to an increase in the viscosity of the nanofluid which causes an increase in friction factor. The effects of Peclet number, Reynolds number, and Nusselt number have been investigated. Those dimensionless number values change with the change in the working fluid viscosity, Prandtl number, and volume concentration of suspended nanoparticles.

4. Conclusion

Advances on studies for improvements on heat exchangers have been performed by several researchers, and in the present book, some of them are presented. These advances are focused on heat transfer enhancement, manufacturing, and working fluids.

Author details

Laura L. Castro Gómez

Address all correspondence to: lauracg@uaem.mx

Centro de Investigación en Ingeniería y Ciencias Aplicadas (CIICAp), Universidad Autónoma del Estado de Morelos (UAEM), Cuernavaca, Mexico

References

[1] Shah RK. Classification of heat exchangers. In: Kakaç S, Bergles AE, Mayinger F, editors. Heat Exchangers: Thermal-Hydraulic Fundamentals and Design. Washington, DC: Hemisphere Publishing; 1981. pp. 9-46

[2] Cengel YA, Ghajar AJ. Heat and Mass Transfer. 4th ed. New York, USA: McGraw Hill; 2011. 921 p

[3] Kakaç S, Liu H, Pramuanjaroenkij A. Heat Exchanger, Selection, Rating and Thermal Design. Boca Raton, EUA: CRC Press, Taylor & Francis Group; 2012 (Chap. 1)

[4] Afgan N, Carvalho M, Bar-Cohen A. In: Butterworth D, Roetzel W, editors. New Developments in Heat Exchangers. New York: Gordon & Breach; 1994

[5] Arab M, Abbas A. Optimization-based design and selection of working fluids for heat transfer: Case study in heat pipes. Industrial & Engineering Chemistry Research. 2014;53(2):920-929. DOI: 10.1021/ie4026709

Design

Chapter 2

Numerical Investigation of PCM Melting in a Finned Tube Thermal Storage

Imen Jmal and Mounir Baccar

Additional information is available at the end of the chapter

http://dx.doi.org/10.5772/intechopen.76890

Abstract

Due to their high energy storage capacity, latent heat storage units using phase change materials (PCMs) have gained considerable attention over the past three decades. The heat exchange of a PCM with the surrounding medium is managed by the thermal energy equation (solidification/melting) with different complex boundary and initial conditions. In this study, we propose to solve numerically this equation applied to a PCM by the finite difference method. To understand the storage phenomenon of solar energy in the form of latent heat in PCM, initially found under cooling at 18°C, we studied the fusion in a specific configuration corresponding to a tubular exchanger with five circular horizontal fins. In this perspective, we propose in this work a numerical investigation based on an enthalpy formulation to study the melting of a PCM in a finned heat exchanger. This numerical approach gives simultaneously the temperature distributions in the PCM storage system and temporal propagation of the melting front during the melting of the PCM when it is exposed to a hot airflow. Also, we give in this study the transient evolution of the longitudinal air temperature profiles.

Keywords: numerical investigation, melting, PCM, latent heat, natural convection

1. Introduction

Every latent heat thermal energy storage (LHTES) should include three main components: an appropriate PCM in the required temperature range, a container for the stocking substance, and an appropriate carrying fluid for an effective heat transfer from the heat source to the heat storage. Moreover, the PCMs require an important heat exchange area because of its low thermal conductivity.

One method is to increase the heat transfer surface area by employing finned surfaces [1–3]. Many numerical and analytical models of PCM solidification and PCM melting in finned PCM-air exchangers have been published in order to evaluate their performance.

Rostamizadeh et al. [4] developed a numerical model of energy storage in a rectangular container of PCM based on an enthalpy formulation, and the effect of PCM thickness on temperature distribution and melting fraction was investigated. Thus, these researchers established that 5 mm is the best thickness of the PCMs. Besides, the study results show that the PCM mass and the melting time verify a linear relationship. Ismail et al. [5] analyzed a numerical and experimental study on the solidification and the melting of PCM around a vertical axially finned isothermal cylinder. The model is based upon the pure conduction mechanism of heat transfer. From the given results, it can be noticed that the fin thickness does not have any important influence on the time of solidification; indeed, the time for full solidification and the solidification rate are strongly affected by the fin length and the number of fins. The difference of temperature has a reverse impact on the solidification of PCM, and the full solidification time tends to decrease when the temperature difference increases. Seck et al. [6] studied the evolution of the melting front of a plate of paraffin (52–54) submitted to sunshine rays in order to design a less voluminous and more efficient heat storage system. The results have shown that the natural convection plays a major role in the kinetics of front propagation. Hence, to numerically reproduce the temporal evolution of the interface throughout the fusion process, the developed natural convective phenomenon should necessarily be taken into account during the melting process. El Qarnia et al. [7] have presented a numerical model of PCM melting in a rectangular cavity heated with three protruding heat sources mounted on a vertical conducting plate. They have carried out numerical investigations for studying the influence of different key parameters on the thermal performance of the PCM-based heat sink. They have developed two correlations, one for the appropriate melt fraction and the other for the maximum working time. The proposed approach can be useful for the design of PCM-based cooling systems.

The present paper presents a numerical investigation of the melting of PCM (paraffin RT27) in a finned tube thermal storage for air conditioning systems taking into account the presence of natural convection. The first part includes the presentation of a numerical model based on the continuity, momentum, and thermal energy equations, treated by the finite volume method. The main objective of this numerical approach is to investigate the temporal evolution of PCM melting and the melting front with the evolution of the liquid fraction, as well as the longitudinal profiles of the heat transfer fluid (HTF) in the duct.

2. Computational domain and mathematical formulation

In the present paper, the studied configuration is a PCM-air heat exchanger with three coaxial cylinders (**Figure 1**). The radius of the inner cylinder is $R_0 = 8$ cm, that of the middle is $R_1 = 13$ cm, and that of the outer is $R_3 = 17$ cm. The thickness of all tubes is 3 mm, and the length of the system is 1.2 m. As shown in **Figure 1**, the space between the inner and middle

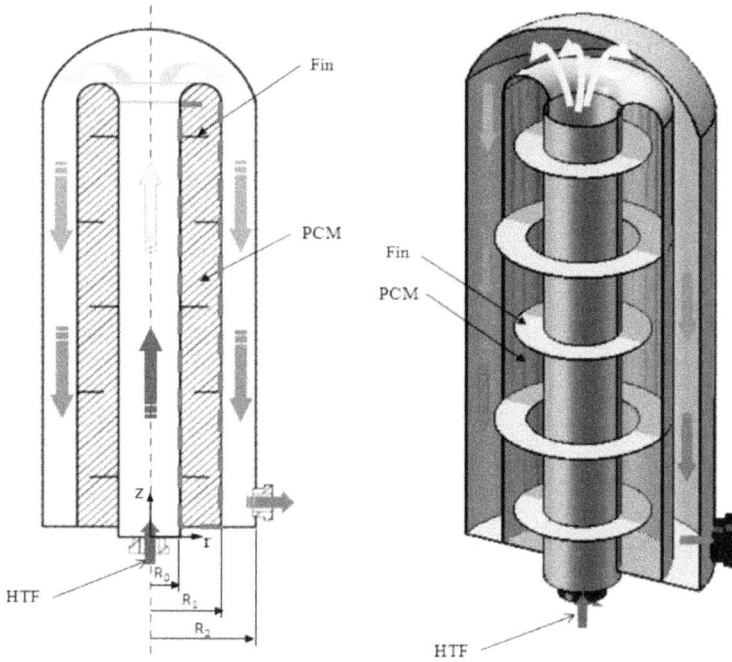

Figure 1. Studied configurations and computational domain.

cylinders is filled with PCM (RT27). In the first-pass flow, the heat transfer fluid (HTF) passes through the central tube, and in the second pass-flow, it flows in the reversal direction in the annular space before exiting the PCM-air heat exchanger module by the same input side. The mass flow rate of air is 0.08 kg.s^{-1}. The fins are made of aluminum with a constant thickness of 3 mm. The thermophysical properties of the solid and liquid PCM are listed in **Table 1**.

The problem is 2D since it has a symmetry of revolution, and hence the computational domain (the hatched area in **Figure 1**) becomes 5 cm in width H and 1.2 m in length L.

2.1. Modeling of the heat transfer to the airflow circling in the axial duct and the annular space

The heat transfer to the HTF-governing equation is given as follows:

$$\rho_{air} \; Cp_{air} \left(\frac{\partial T_{air}}{\partial t} + w_{air} \frac{\partial T_{air}}{\partial z} \right) = S_T \tag{1}$$

For the first passage of air,

$$S_T = \frac{2 \; h_i \; (T_{PCM} - T_{air})}{R_0} \Big/ h_i = 166 \text{ W m}^{-2°} \text{C}^{-1} \tag{2}$$

		RT27
Melting interval [°C]		[27–27,5]
Latent heat [kJ.kg^{-1}]		179
Density [kg.m^{-3}]	Solid	870
	Liquid	760
Specific heat [kJ.kg^{-1}.°C^{-1}]	Solid	2.4
	Liquid	1.8
Thermal conductivity [W.m^{-1}.°C^{-1}]	Solid	0.24
	Liquid	0.15
Thermal volumetric expansion coefficient [kg.m^{-3}.°C^{-1}]		8.5×10^{-4}
Kinematic viscosity [m^2.s^{-1}]		1.5×10^{-6}

Table 1. Thermophysical properties of the PCM RT27 [8].

For the second passage of air,

$$S_T = \frac{2 \ R_1 \ h_o \ (T_{PCM} - T_{air})}{R_2^2 - R_1^2} / h_o = 114 \ \text{W m}^{-2°}\text{C}^{-1} \tag{3}$$

2.2. Numerical modeling of heat transfer in the PCM domain

The natural convection of a transient laminar PCM flow in the PCM container has been studied numerically. Considering constant properties, except the density difference (Boussinesq approximation), and using the enthalpy method, the continuity, momentum, and thermal energy equations for the PCM can be written as follows:

2.2.1. Continuity equation

$$\frac{1}{r}\frac{\partial(ru)}{\partial r} + \frac{\partial w}{\partial z} = 0 \tag{4}$$

2.2.2. Momentum equations

The momentum equations incorporate sink terms to take into account the changing phase of the PCM. The condition that all velocities in solid regions are zero is considered using an enthalpy-porosity approach [9].

$$\frac{\partial U}{\partial t} = - \ \text{div} \ \left(\vec{V} \ U - v \ \vec{\text{grad}} \ U \right) - \frac{1}{\rho}\frac{\partial P}{\partial r} - v \ \frac{1}{r}\left(\frac{2}{r}\frac{\partial V}{\partial \theta} + \frac{U}{r} \right) + \frac{V^2}{r} - \frac{1}{\rho}c \ \frac{(1-f)^2}{f^3 + b} \ U \tag{5}$$

$$\frac{\partial W}{\partial t} = - \ \text{div} \ \left(\vec{V} \ W - v \ \vec{\text{grad}} \ W \right) - \frac{1}{\rho}\frac{\partial P}{\partial z} + g \ \beta \ (T - T_0) - \frac{1}{\rho}c \ \frac{(1-f)^2}{f^3 + b} \ W \tag{6}$$

where S_u is constituted by two terms: the first of which corresponds to the melting sink term, and the second is the thermo-convective generation term. Furthermore, S_w is the extinction of "W" component due to the melting of PCM.

2.2.3. Energy equation

$$\rho(T)\ C_p(T)\ \frac{\partial T}{\partial t} = -\text{div}\left(\rho(T)\ C_p(T)\ \vec{V}\ T - \lambda(T)\ \vec{\text{grad}}\ T\right) \qquad (7)$$

where [10]

$$\rho(T)Cp(T) = \begin{cases} \rho_s Cp_s & T \leq T_{sd} \\ \rho_l Cp_l & T \geq T_{liq} \\ \dfrac{\rho_{mix}\ L}{T_{liq} - T_{sd}} & T_{sd}\ \leq T \leq T_{liq}/\rho_{mix} = f\rho_l + (1-f)\rho_s \end{cases} \qquad (8)$$

$$\lambda(T) = \begin{cases} \lambda_s & T \leq T_{sd} \\ \lambda_l & T \geq T_{liq} \\ \dfrac{\lambda_l - \lambda_s}{T_{liq} - T_{sd}}\left(T - T_{liq}\right) + \lambda_l & T_{sd} \leq T \leq T_{liq} \end{cases} \qquad (9)$$

The liquid fraction "f" can be analyzed by the following equation:

$$f = \frac{T - T_{sd}}{T_{liq} - T_{sd}} \qquad (10)$$

2.3. Initial and boundary conditions

For the velocity components, the nonslip boundary conditions on all solid walls are imposed.

The thermal limiting condition imposed on the interfaces between the PCM envelope and the air evolving in the tubular exchanger results from the application of the continuity rule. Thus, a conducto-convective mixed condition is applied on both sides of the PCM container. During the first passage of air, the heat transfer between the PCM and the HTF through the wall of the PCM container is written as follows:

$$h_i(T_{PCM} - T_{air}(z)) = \lambda \frac{\partial T}{\partial r}\bigg|_{r=R_0} \qquad (11)$$

During the second air passage,

$$h_o(T_{PCM} - T_{air}(z)) = -\lambda \frac{\partial T}{\partial r}\bigg|_{r=R_1} \qquad (12)$$

As initial conditions, it is assumed in the charging mode (melting) that the storage unit and the air in the duct are initially at a uniform temperature of 45°C and the temperature of the inlet air is 18°C.

2.4. Resolution method

In the present study, a finite volume method for the numerical solution of 2D unsteady natural convection flow and energy Eqs. (5)–(7) is used [1]. Thus, the domain including the PCM is divided into a convenient number of control volumes (NR = 40; NZ = 300).

The staggered grid method is employed to accurately control mass conservation and heat transfer on all control volumes, and an exponentially fitted spatial discretization scheme is used. Furthermore, the equation is solved by the alternating direction implicit (ADI) technique, in which the Crank-Nicolson scheme is used for the time discretization, and the SIMPLER algorithm for the treatment of the pressure-velocity coupling.

Numerical results are obtained by developing a specific code using the FORTRAN language.

3. Validation of the numerical approach

The performance of the proposed method is verified with the experimental data performed by Longeon et al. [11]. This was done by simulating, under the same operating conditions, the melting process of the paraffin RT35 used as PCM, with the thermophysical properties as listed in **Table 2**.

The PCM storage system is composed of two concentric cylinders (**Figure 2**): the first one with an inner diameter of 44 mm is made of Plexiglas, and the other with an inner diameter of 15 mm is made of stainless steel. The length of the whole system is 400 mm. The HTF flows into the inner tube, and the PCM fills up the annular space. As it is shown in **Figure 2**, the thermocouples are distributed on section B.

It is found that the temperature evolution does not depend on angular position, indicating that the heat exchange can be modeled with a 2D approach. In **Figure 3**, which gives the temperature evolution with time at two radial positions (a and b), our numerical results are compared with those determined experimentally by Longeon et al. [11]. A good agreement between numerical and experimental data can be noted. In fact, in both positions, the same shape tendency with a rapid decrease of the temperature at the beginning followed by a slow rating

Melting point [°C]		35
Latent heat [kJ.kg^{-1}]		157
Density [kg.m^{-3}]	Solid	880
	Liquid	760
Specific heat [kJ.kg^{-1}.K^{-1}]	Solid	1.8
	Liquid	2.4
Thermal conductivity [W.m^{-1}.K^{-1}]		0.2
Kinematic viscosity [m^2.s^{-1}]		3.3×10^{-6}

Table 2. Thermophysical properties of RT35 [11].

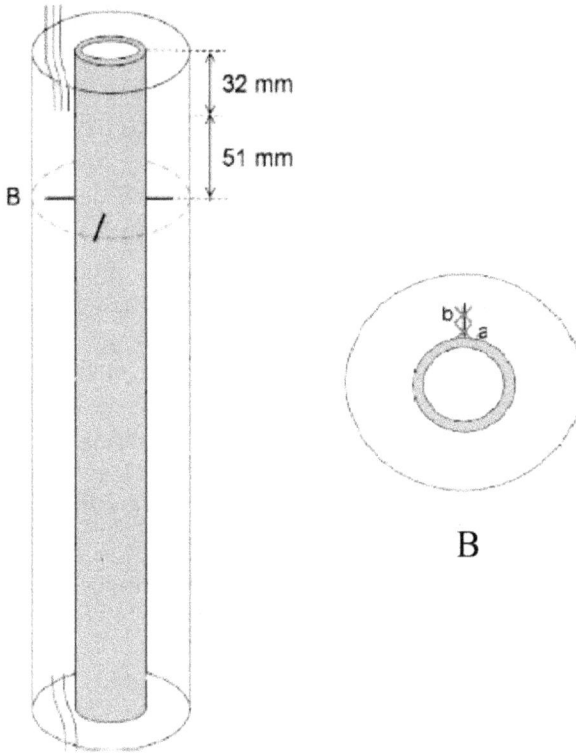

Figure 2. Scheme of the experimental loop and thermocouple positions in the test section [11].

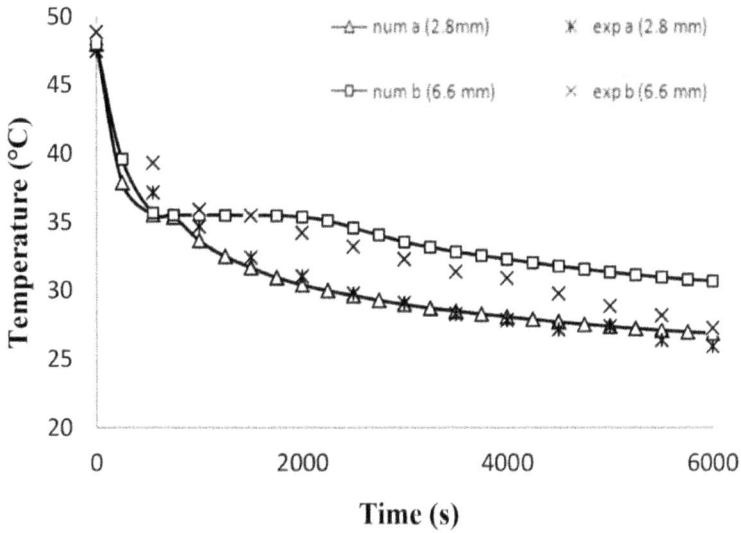

Figure 3. Comparison of the experimental results and our numerical solution.

decrease characterizing the phase change process can be noted. The average deviation between our numerical simulation and the experimental data of Longeon et al. [11] is approximately 5%. This deviation is very acceptable for this type of study [12].

4. Results

4.1. Temporal evolution of temperature and velocity fields

Figures 4 and **5** present the distinctive temporal evolution of the distribution of the temperature and current lines. The analysis of the thermal fields has shown that the temperature is higher in the inner part near the central tube and the fins. The peripheral part is also warm, but to a lesser degree, because this surface is exposed to less warm air, since it has lost some of its sensible heat during the first passage. In the early stages of the fusion, very little free space exists for the flow. Hence, the homologous mappings given the current lines and the isotherms have revealed that the liquefaction front is parallel to the tubes and the fins, which confirms the conduction predominance.

As a result of the free convection movements which have become increasingly strong in molten paraffin, the volume is mainly extended over the entire height of the warmer central tube.

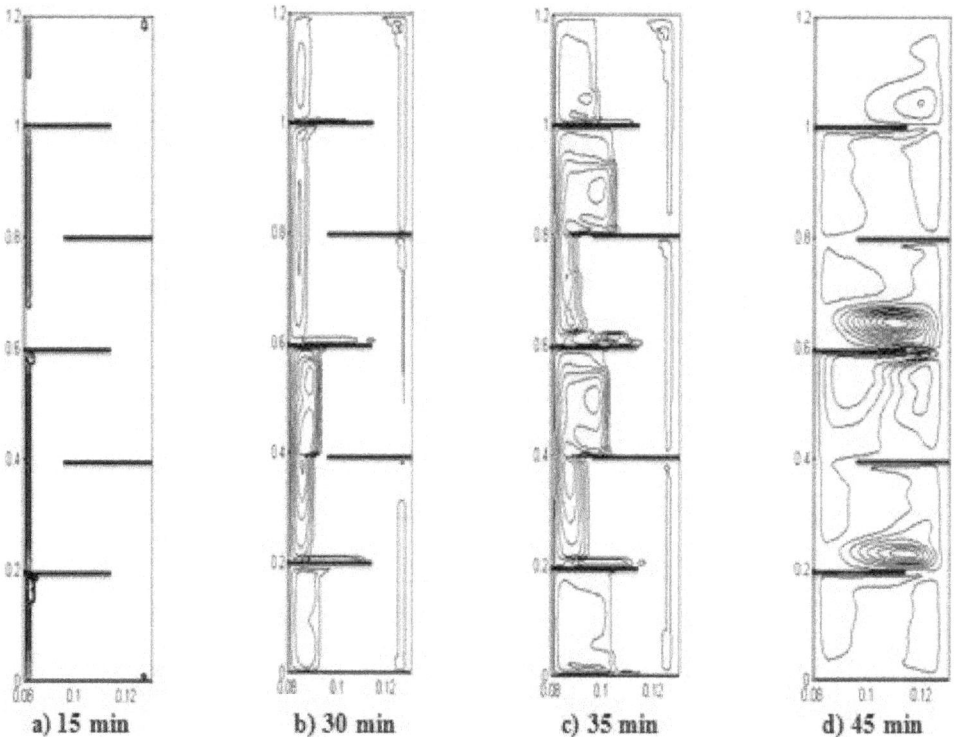

a) 15 min   b) 30 min   c) 35 min   d) 45 min

Figure 4. Temporal evolution of the temperature field and with corresponding current lines.

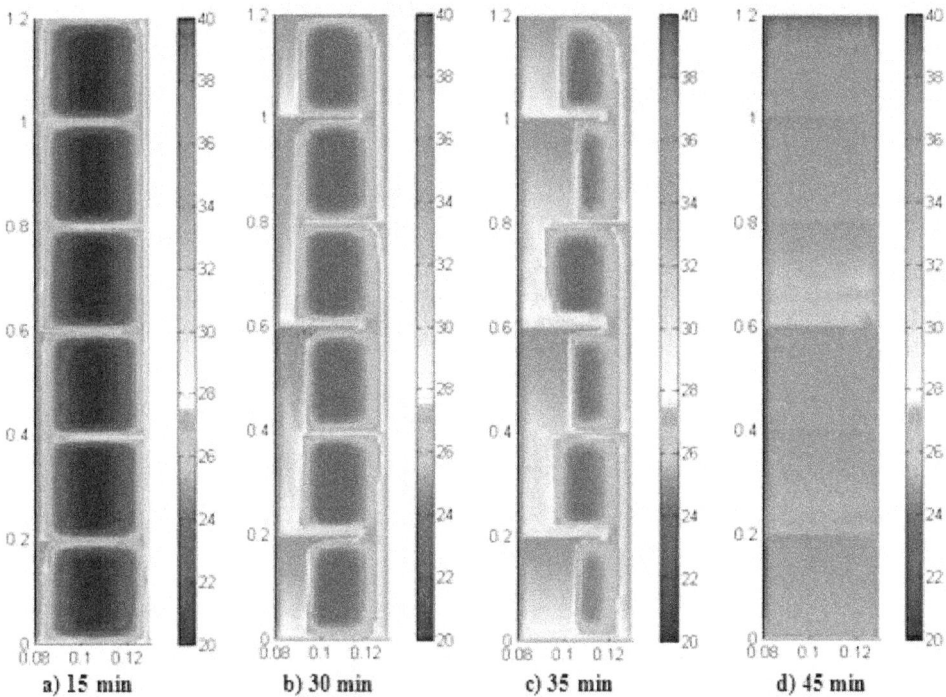

Figure 5. Temporal evolution of temperature fields in heat storage mode (five circular fins).

While mounting along the air-heated tube, the molten paraffin progressively accumulates the heat of the central tube and comes against the fins before diverting toward the interior of the domain. Then, it descends along the solid-liquid interface by gradually releasing the accumulated heat, till reaching the bottom of the liquefied PCM cavity. The same phenomenon is observed on the temperature profiles near the outer tube. The observation of the temperature evolution in the figure reveals that the conduction phenomenon predominates on the opposite side of the tubes.

For all heat storage times, a comparison of the temperature levels in the zones located between two consecutive internal fins shows higher temperature levels in the upper zones.

Figure 6 shows the longitudinal profiles of the air evolving in the exchanger. In storage mode, the air entering the exchanger at 45°C is cooled in the first passage and then in the second passage by providing its PCM-sensible heat, initially at 18°C. This gradually leads to:

- preheating the PCM to 18°C at its melting point (27°C);

- liquefying the PCM in the temperature range 27–27.5°C; and

- overheating the liquefied PCM.

These exchange natures are presented in **Figure 7**, in which the temperature evolution values of the air leaving the exchanger as a function of time are plotted. Indeed, in the time interval

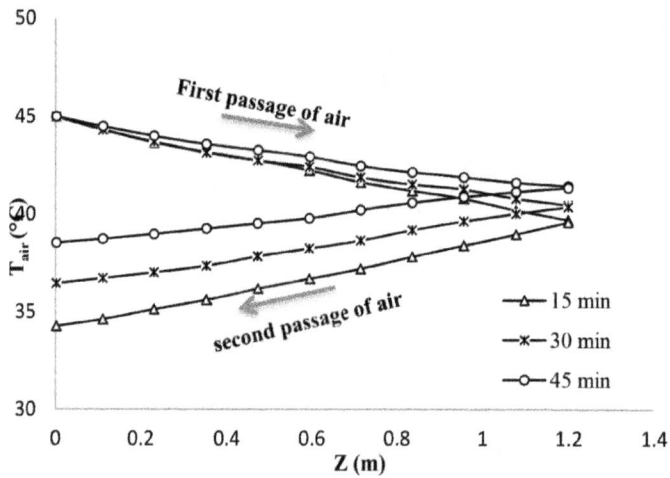

Figure 6. Temporal evolution of the longitudinal air profile in the duct.

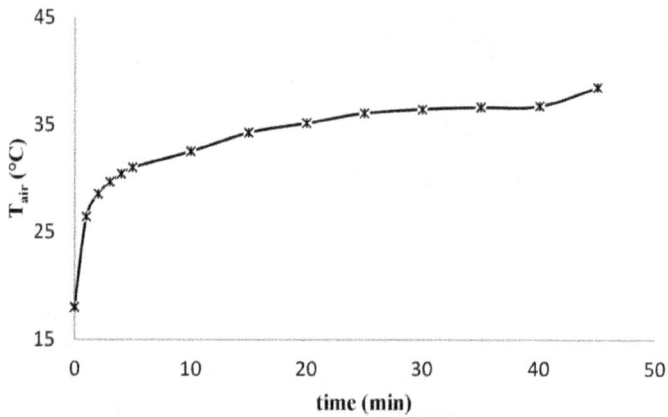

Figure 7. Temporal evolution of the air temperature at the outlet air duct.

between 5 and 40 min, an almost constant output temperature can be observed, indicating a heat supply that is equally constant during the phase of state change.

4.2. Temporal evolution of the melting front

Figure 8 shows the temporal evolution of the melting front during the energy storage. After a very short time (about 1 min), we have already observed that the fusion initiates at the wall of the central tube, heated to a temperature higher than that of the material liquefaction by the HTF.

During the PCM melting, three successive transfer regimes can be noted in the transmission mode of the heat: in the first moments, the zone occupied by the liquid phase has a very small width, so there is practically no convection heat transfer. Consequently, the liquid-solid interface moves parallel to the tubes and the fins under the unique action of conduction. At this stage, the morphology of the melting/fusion front is presented as a flat surface. Then, when the width of the liquid phase gives birth to convection, the regime is called mixed and corresponds to a combination of natural convection and conduction. This results in a higher melting rate in the solid part; the shape of the interface is thus modified. Finally, when the liquid phase has become wide enough to allow a full development of the convective movements, the convective regime is reached, in which the main part of the heat transfer inside the liquid phase is due to the natural convection. The solid-liquid interface then is slightly deformed, and the natural convection will be the main motor for the enhancement of the melting front inside the PCM enclosure.

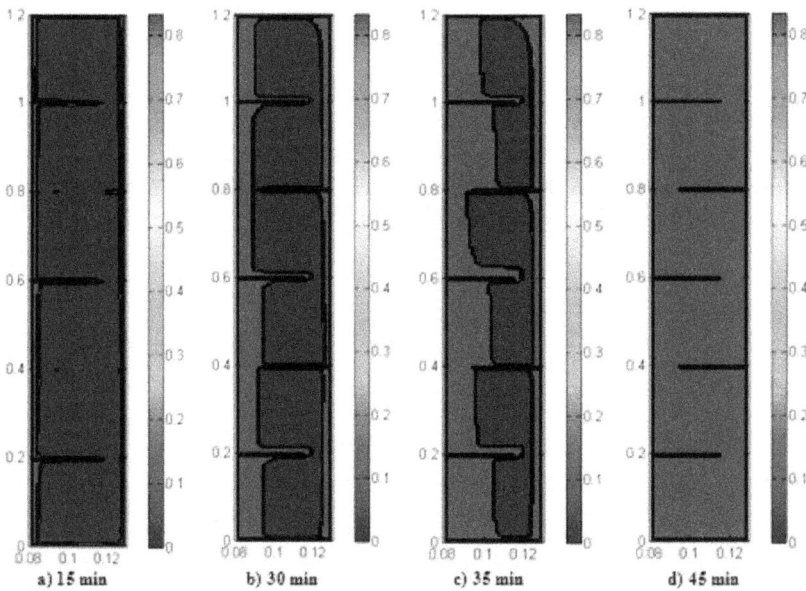

Figure 8. Temporal evolution of the melting front of the storage mode.

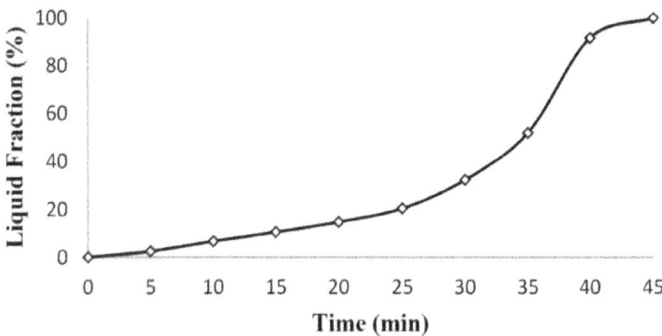

Figure 9. Transient evolution of liquid fraction.

These transfer regimes are clearly seen in **Figure 9**, in which the evolution of the liquid fraction is shown, indicating the melted volume according to the total volume, as a function of time. Initially, there was a low growth rate corresponding to the conductive regime. Between roughly 20 min and 30 min, the growth rate becomes more important, underlining the contribution of thermo-convective transfer. Starting from 35 min, the liquefaction rate becomes more important, highlighting the dominance of the thermo-convective effects. The total liquefaction ends after 45 min.

5. Conclusion

In the present work, we have studied the melting of the PCM (Paraffin RT27) in a tubular PCM-air heat exchanger. We have concluded that in the storage mode, the complete PCM melting time was found to be substantially shorter (about one-fourth of the time) compared to that recorded for solidification during the destocking [10]. Indeed, in contrast to what happens during solidification involving an increase in the thermal resistance to the wall due to the formation of a solid PCM layer, the initiation of liquefaction on the side of the hot walls promotes the formation of thermo-convective loops which are intensified during the fusion.

Author details

Imen Jmal* and Mounir Baccar

*Address all correspondence to: imenn.jmal@gmail.com

Research Unit of Computational Fluid Dynamics and Transfer Phenomena (CFDTP), Mechanical Engineering Department, National Engineering School of Sfax, Tunisia

References

[1] Al-Abidi Abduljalil A, Sohif Mat Sopian K, Sulaiman MY. Numerical study of PCM solidification in a triplex tube heat exchanger with internal and external fins. International Journal of Heat and Mass Transfer. 2013;**61**:684-695

[2] Bauer T. Approximate analytical solutions for the solidification of PCMs in fin geometries using effective thermo-physical properties. International Journal of Heat and Mass Transfer. 2011;**54**:4923-4930

[3] Mosaffa AH, Talati F, Basirat Tabrizib H, Rosen MA. Analytical modeling of PCM solidification in a shell and tube finned thermal storage for air conditioning systems. Energy and Buildings. 2012;**49**:356-361

[4] Rostamizadeh M, Khanlarkhani M, Mojtaba Sadrameli S. Simulation of energy storage system with phase change material (PCM). Energy and Buildings. 2012;**49**:419-422

[5] Ismail KAR, Alves CLF, Modesto MS. Numerical and experimental study on the solidification of PCM around a vertical axially finned isothermal cylinder. Applied Thermal Engineering. 2001;**21**:53-77

[6] Seck D, Thiam A, Sambou V, Azilinon D, Adj M. Détermination du front de fusion d'une plaque de paraffine soumise à l'ensoleillement. Journal des Sciences. 2009;**1**:34-42

[7] El Qarnia H, Draoui A, Lakhal EK. Computation of melting with natural convection inside a rectangular enclosure heated by discrete protruding heat sources. Applied Mathematical Modelling. 2013;**37**:3968-3981

[8] Aadmi M, Karkri M, El Hammouti M. Heat transfer characteristics of thermal energy storage for PCM (phase change material) melting in horizontal tube: Numerical and experimental investigations. Energy. 2015:1-14

[9] Brent AD, Voller VR, Reid KJ. Enthalpy-porosity technique for modeling convection-diffusion phase change: Application to the melting of a pure metal. Numerical Heat Transfer. 1988;**13**:297-318

[10] Jmal I, Baccar M. Numerical study of PCM solidification in a finned tube thermal storage including natural convection. Applied Thermal Engineering. 2015;**84**:320-330

[11] Longeon M, Soupart A, Fourmigué JF, Bruch A, Marty P. Experimental and numerical study of annular PCM storage in the presence of natural convection. Applied Energy. 2013;**112**:175-184

[12] Oberkampf WL, Barone MF. Measures of agreement between computation and experiment: Validation metrics. Journal of Computational Physics. 2006;**217**:5-36

Chapter 3

Use of Heat Transfer Enhancement Techniques in the Design of Heat Exchangers

Martín Picón-Núñez, Jorge C. Melo-González and
Jorge Luis García-Castillo

Additional information is available at the end of the chapter

http://dx.doi.org/10.5772/intechopen.78953

Abstract

Heat transfer enhancement refers to application of basic concepts of heat transfer processes to improve the rate of heat removal or deposition on a surface. In the flow of a clean fluid through the tube of a heat exchanger, the boundary layer theorem establishes that a laminar sublayer exists where the fluid velocity is minimal. Heat transfer through this stagnant layer is mainly dominated by thermal conduction, becoming the major resistance to heat transfer. From an engineering point of view, heat transfer can be enhanced if this stagnant layer is partially removed or eliminated. In single-phase heat transfer processes, three options are available to increase the heat transfer rate. One of them is the choice of smaller free flow sectional area for increased fluid velocity bringing about a reduction of the thickness of the laminar sublayer. A second option is the engineering of new surfaces which cause increased local turbulence, and the third option consists in the use of mechanical inserts that promote local turbulence. The application of these alternatives is limited by the pressure drop. This chapter describes the concept of heat transfer enhancement and the ways it is applied to the development of new heat exchanger technology.

Keywords: heat transfer enhancement, compact surfaces, turbulence promoters, pressure drop, thermohydraulic performance

1. Introduction

Growing interest in thermal energy recovery in the process industry has driven the development of new heat exchanger technology by means of heat transfer enhancement techniques. Heat transfer enhancement techniques can be as basic as the manipulation of the fluid velocity inside the unit or as complex as the design of new surface geometries or the design of inserts in the case of tubular geometries.

One of common classifications of heat exchangers is based on the level of compactness that refers to how much heat transfer surface area (m²) a heat exchanger fits within a unit volume (m³). In this regard, heat exchangers are classified into compact and noncompact. For instance, tubular heat exchangers such as the shell and tube type are considered within the non-compact technology.

Compact heat exchangers are characterized by a high level of heat transfer enhancement due to their geometrical features and particularly the shape of their heat transfer surfaces that maximize the heat transfer rate by the generation of local turbulence at the expense of increased pressure drop [1]. On the other hand, heat transfer enhancement techniques applied to tubular heat exchangers seek to promote local turbulence by mechanical means [2].

From the thermohydraulic point of view, the main objectives that new exchanger technology must fulfill are:

1. Smaller required heat transfer area and volume for the same heat duty and pressure drop consumption.

2. Increased heat load for the same installed heat transfer area within the limitations imposed by the permitted pressure drop.

For these objectives to be met, it follows that any heat transfer enhancement technique must improve on the heat transfer coefficient and bring about a reduction in the heat transfer area for the same heat duty in a new design. Alternatively, in existing exchangers, enhanced techniques will increase their heat transfer capacity. Heat transfer enhancement techniques are classified into two main groups: active and passive. Active methods require external power, for instance, fluid suction or injection, surface fluid vibrations, etc. Passive methods consist in the modification of the heat transfer surface of the system. The main feature of such devices is that they reduce the laminar boundary layer next to the walls which is the major resistance to heat transfer. In the case of tubular exchangers, passive enhancing methods involve the use of turbulence promoters. Due to its effectiveness, low cost, simple installation and removal for cleaning, availability in almost any material of construction and reliability, such option has become a key heat transfer enhancement technique in recent years. Additionally, tube inserts have proved effective in fouling mitigation in applications with fluids with a high tendency to foul.

This chapter will focus on the various ways heat transfer is enhanced and gives birth to new exchanger technology, either in the way of new compact exchanger technology or in the way of internal inserts to be used in conventional tubular technology.

2. Fluid velocity and heat transfer enhancement

In heat exchanger design or retrofit, fluid velocity can be increased through the choice of reduced free flow area or sectional area. Another option available in design is the choice of the number of passes. In an existing unit, any change to the geometry is referred to as retrofit.

As mentioned earlier, the increase of fluid velocity brings about a reduction of the thickness of the laminar sublayer which in turn results in increased heat transfer coefficients. For the case of tubular heat exchangers, the heat transfer coefficient on the tube and shell side follow the expression of the form:

$$h_t = K_t \, v_t^{0.8} \tag{1}$$

$$h_s = K_s \, v_s^{0.6} \tag{2}$$

where ht and hs are the tube side and shell side heat transfer coefficients; vt and vs are the tube side and shell side mass flow rates; Kt and Ks are parameters that involve geometrical features and physical properties. The change in heat transfer coefficient with increased velocity for the tube side can be calculated from:

$$\frac{h_t^N}{h_t^0} = (F_t)^{0.8} \tag{3}$$

$$F_t = \frac{v_t^N}{v_t^0} \tag{4}$$

where v_t^N and v_t^0 are the fluid velocities at the new and original conditions, respectively. A similar expression can be derived for the case of the shell side. The thermal behavior of the exchanger with velocity is limited by the hydraulic behavior of the unit represented by the increase of the pressure drop. It can be demonstrated that for the tube side, the way pressure drop varies with velocity is expressed as:

$$\Delta P_t = K_{pt} \, v_t^{1.9} \tag{5}$$

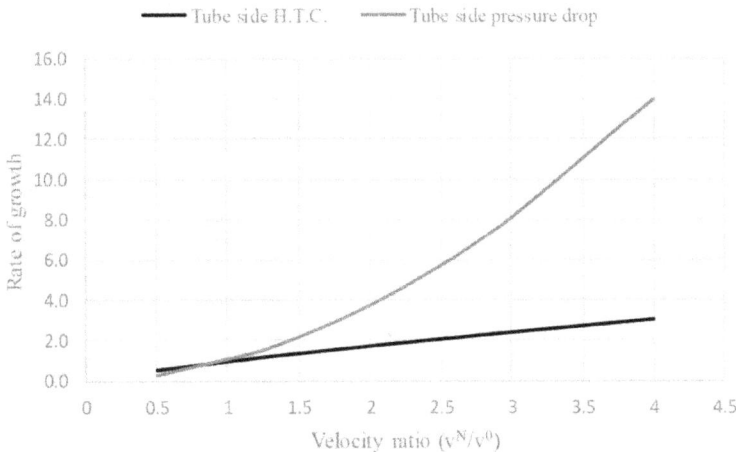

Figure 1. Rate of growth of the heat transfer coefficient and pressure drop on the tube side for a range of velocity increase ratios.

Therefore, the change in pressure drop with velocity can be approximated by:

$$\frac{\Delta P_t^N}{\Delta P_t^0} = (F_t)^{1.9}$$

(6)

where ΔPt^N and ΔPt^0 are the new pressure drop and the original pressure drop. Expressions (3) and (6) indicate that the pressure drop grows faster with velocity than the heat transfer coefficient does as shown in **Figure 1**.

Although velocity is a simple way of enhancing the heat transfer performance of a heat exchanger, the main problem associated with its manipulation is that the rate at which pressure drop grows establishes a limit. Therefore, in any retrofit or design approach, the allowable pressure drop determines how much heat transfer intensification can be achieved.

3. Heat transfer surface design

For the same bulk fluid velocity, the surface geometry becomes an alternative design option for improving the thermohydraulic performance of a heat exchanger. Examples of modified surfaces used in tubular heat exchangers are the twisted tube exchanger (**Figure 2**). The swirl flow motion creates a constant removal of the laminar layer, thus increasing the heat transfer coefficient with minimal increase of the pressure drop. This geometry involves both the internal and the external fluid.

In operation, the shell side of conventional heat exchangers represents an important area of opportunity for improvement. Heat transfer dampening due to the creation of stagnant zones and high pressure drop due to abrupt changes of direction can be avoided using helical baffles. Helical baffle heat exchangers exhibit a more uniform flow distribution on the shell side compared to conventional shell and tube exchangers for the same pressure drop [3]. They can also reduce tube vibrations and fouling. Their manufacturing costs are higher, but they require less maintenance and their operating costs are lower which so in the long term the investment pays-off [4]. The geometrical features (**Figure 3**) that define this type of technology

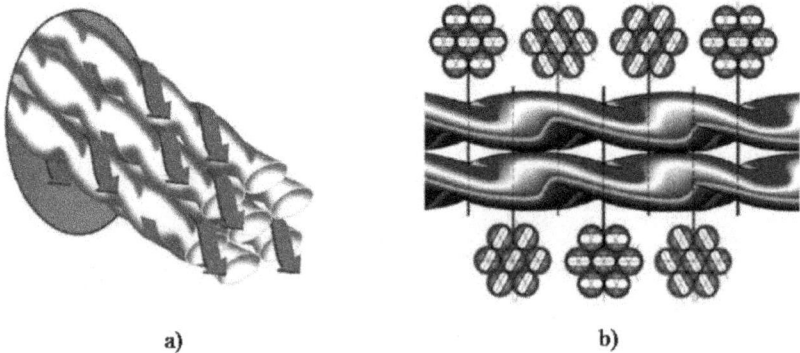

a)   b)

Figure 2. (a) Twisted tube heat exchanger technology; and (b) twisted tube construction.

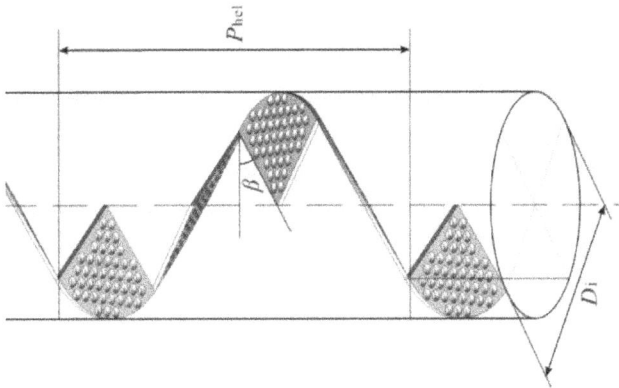

Figure 3. Main geometrical features of a helical heat exchanger.

are: helical pitch, Phel (distance between two consecutive baffles); the helical angle, β (the angle formed between the helix and the vertical); and the shell diameter, Di, [5].

In [5], it has been demonstrated that the highest thermal performance in this type of geometry is achieved when the baffle angle is 40°. Numerical studies in 3D [3] indicate that the heat transfer rates in helical baffle exchangers are higher than in conventional segmental baffle exchangers from 9 to 23%. Other design alternatives include the use of multiple shell passes [6]. The improved performance of these types of units has also been demonstrated experimentally as reported in [5]. There is little information about expressions to estimate the thermohydraulic performance of helical exchangers and the few expressions available are reported for ideal flow conditions and for design purposes correction factor must be applied. In recent work [7], heat transfer and friction factor expressions derived from experimental data have been reported for different helix angles.

Helical-baffle heat exchangers exhibit superior performance compared to conventional segmental baffle exchangers. In [8], a short-cut design approach for helical baffle heat exchangers based on the concept of full use of available pressure drop was developed. The application of this design methodology indicates that improved designs are possible with helical baffle exchangers. For the same heat duty and pressure drop consumption, a reduction of almost 30% in surface area is obtained.

In the search for improved thermal surface performance, heat exchangers have evolved to what are called compact heat exchangers. A large variety of new surface designs are available. The main feature of these technologies is that for the same pressure drop, they create higher heat transfer coefficients. Some of the most common types of compact heat exchangers are described below.

3.1. Plate and fin heat exchangers

A plate and fin heat exchanger is a type compact heat exchanger that is mainly used in gas to gas applications. It consists of a stack of alternate plates called parting sheets and corrugated fins brazed together as a block as shown in **Figure 4**.

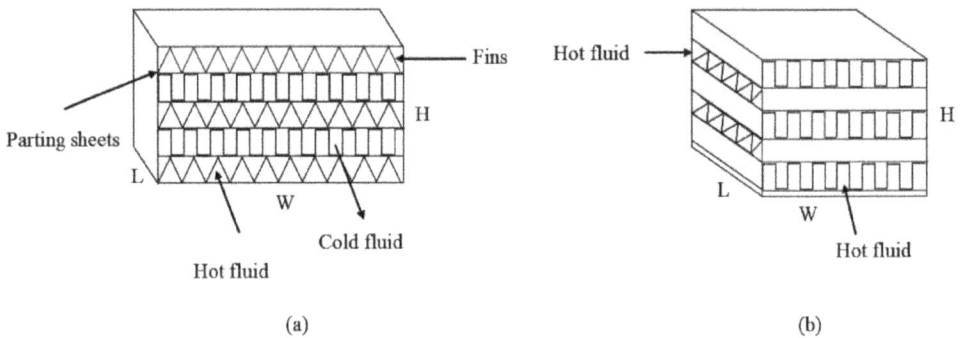

Figure 4. Plate and fin heat exchanger: (a) counter flow arrangement and (b) cross flow arrangement.

Streams exchange heat as they flow through the passages created between the fins and the parting sheets. The fins serve two main functions: a thermal function by increasing the total surface area for heat transfer and a mechanical function by providing mechanical support between layers.

Overall, the design of a heat exchanger requires the specification of the heat duty, stream allowable pressure drops, and certain aspects of exchanger geometry. In the case of plate and fin exchangers, it is fundamental to define the type of the specific secondary surfaces to be employed along with its geometrical features.

The principal geometrical features of a plate and fin exchanger are: ratio of total surface area of one side of the exchanger to volume between plates (βs), plate spacing (δ), ratio of secondary surface area to total surface area (fs), hydraulic diameter (dh), fin thickness (τ), and fin thermal conductivity (k). There are several different types of fins available for design [9]. Among them are: (a) plain-fin, (b) perforated-fin, (c) offset-fin, and (d) wavy (**Figure 5**).

3.2. Plate and frame heat exchangers

Plate and frame heat exchangers (PFHE) are becoming a suitable alternative to shell and tube exchangers in some applications in the process industries. Their construction with characteristics that facilitate the increase of surface area and their versatility in terms of the materials of construction are some of the reason why in some applications they are the best option. This technology encounters applicability limitations in situations with high pressure and temperature exist or and in cases with a large difference in operating pressure between streams since this can cause plate deformations [1, 10]. This technology is also limited in cases with dirty fluids as their small free flow area is prone to passages clogging.

Plate and frame heat exchangers are formed by a series of corrugated plates that are stacked in a frame; one end of the frame is fixed and the other end is movable to allow the addition or removal of plates. The plates are mounted on the frame by means of upper and lower guiding bars and supported by fastening bolts. The space between plates is sealed by means of polymer gaskets. With PFHE, the need for distribution headers is eliminated since ports are an essential element of the plate design and are incorporated into it. The geometry of a

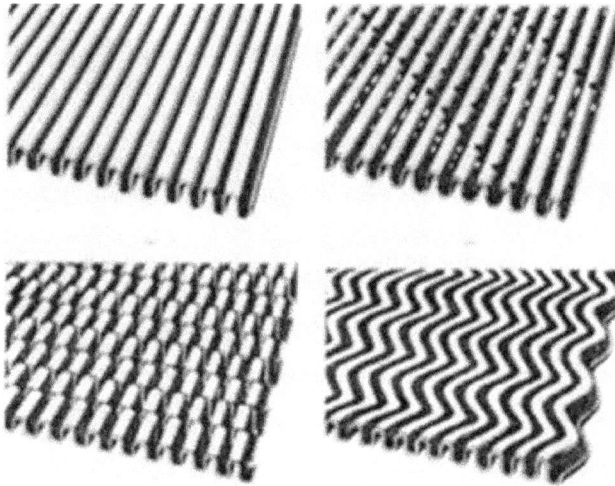

Figure 5. Some of the available secondary surfaces for plate-fin heat exchangers: (a) top left: plain-fin, (b) top right: perforated-fin, and (c) bottom left: offset-fin, (d) bottom right: wavy.

PFHE is characterized by: number of plates (N), number of passages (P), plate length (L), plate width (W), chevron angle (β), plate spacing (b), port diameter (Dp), plate thickness (τ), and enlargement factor (ϕ). The latter term refers to the ratio of the actual surface area of a plate to the area projected on the plane. **Figure 6** shows a typical exchanger assembly and the geometrical features [11].

The corrugation of the plates in a PFHE strongly determines the performance of the unit. The most common corrugation is the chevron type. This is characterized by an angle β with

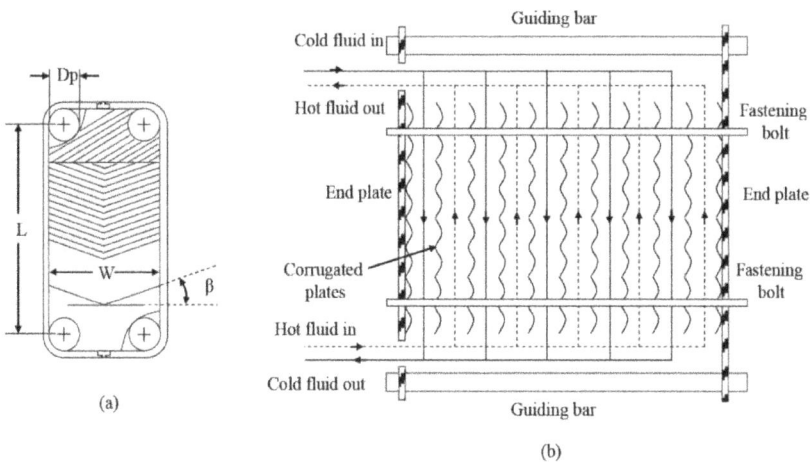

Figure 6. Plate and frame heat exchanger: (a) geometry of a chevron plate and (b) overall exchanger assembly.

respect to the horizontal perpendicular to the direction of the flow. Designs with a low β angles present high level of turbulence, high heat transfer coefficients, and pressure drop, whereas designs with high values of β exhibit low turbulence and therefore lower heat transfer coefficients and pressure drops. The chevron angle is depicted in **Figure 6(a)**.

The flexible construction of PFHE is of great value during retrofit either for increased throughput or for increased heat recovery. The structure can easily be adjusted to achieve the required heat load within the restrictions of the specified pressure drop. This can be achieved by: (a) increasing or reducing the number of thermal plates, (b) changing the type of plate, or (c) by modifying the number of passes.

There are several possible flow pass arrangement options with this type of exchangers; however, all of them stem from the combination of the three basic types (**Figure 7**): series, circuit, and complex. If the variable P represents the number of passes, the number of thermal plates N in a series arrangement is N = 2P − 1. A series arrangement is the choice in situations with low flow rates and with close temperature approaches. Circuit arrangements are more commonly used in applications with large flow rates and with close temperature approaches. Multiple pass arrangements result from the combination of the circuit and series arrangement; they are used when higher velocities for increased absorption of pressure drop is desirable.

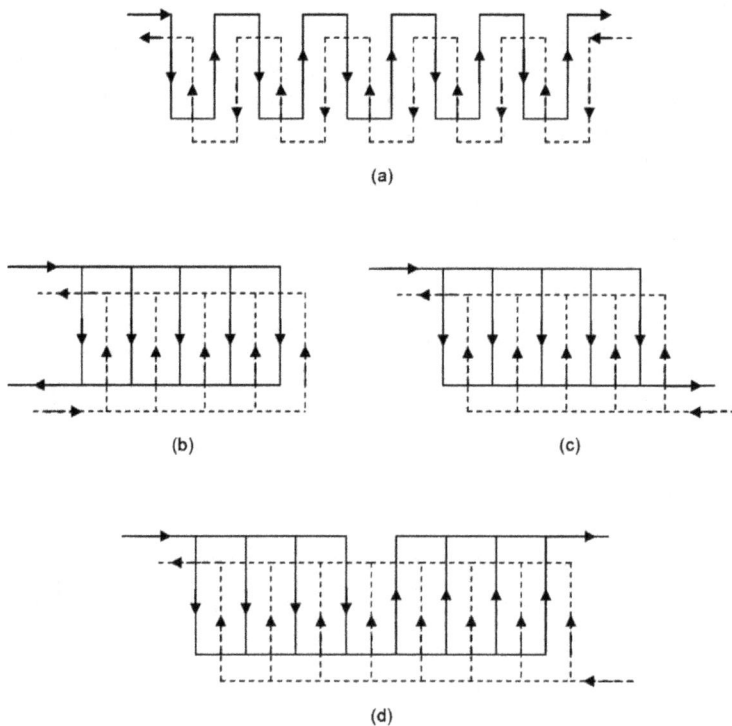

(a)

(b)   (c)

(d)

Figure 7. Flow passage arrangements in plate and frame heat exchangers: (a) series, (b) U circuit, (c) Z circuit, and (d) single pass-two pass.

Fundamental aspects in the design of a PFHE are the thermo-hydraulic performance of the plates and its size. They are a degree of freedom when sizing the unit to meet the given heat load subject to the restrictions of pressure drops of the fluids [12].

From the operating point of view, there are two main situations that reduce the thermal effectiveness of PFHE: (a) flow maldistribution and (b) the effect of end channels and channels between passes. The problem of flow maldistribution arises because of unequal pressure drop in channels [13]. On the other hand, the end plate effect is the unavoidable result of the inherent geometry of this type of technology [14].

3.3. Spiral heat exchangers

Spiral heat exchangers (SHE) are suitable for applications that involve highly viscous liquids and dirty fluids. In a spiral heat exchanger, the fluid is forced to a continual change of direction it flows through the unit as shown in **Figure 8**. This motion change creates high shear stress that eliminates stagnant zones and increases heat transfer coefficients compared to conventional flat surfaces. The continual change of direction also maintains suspended solids in motion preventing their deposition and thus reducing fouling [15].

Even though the fluids flow is overall counter-current arrangement, careful analysis of the local temperature driving forces indicates that this geometry creates disturbance on the driving forces making it depart from an ideal counter-current behavior. The operation of a SHE is associated to two thermal situations that reduce its effectiveness. One is related to what is called the end effects. For instance, all along the length of the unit, each stream exchanges heat with two adjacent streams but in the innermost and outermost part of the unit, heat is transferred only to one side of the channel as shown in **Figure 8**. The second thermal effect is related to the exchange of heat with two adjacent streams. In internal channels, the hot fluid exchanges heat across two adjacent cold channels each at different temperatures; this situation results in disturbed temperature driving forces. These effects are accounted for in design by means of correction factors. The determination of the correction factor for this geometry is

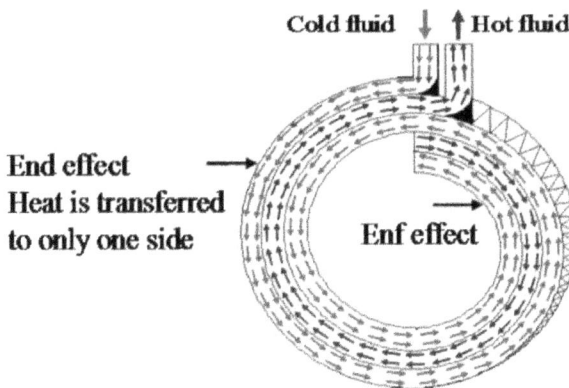

Figure 8. Relative flow of fluids within a spiral heat exchanger.

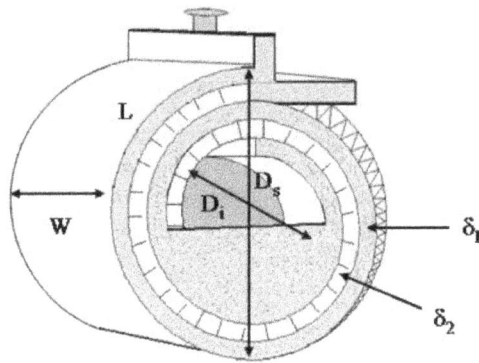

Figure 9. Geometrical features of a spiral heat exchanger.

quite elaborate; however, there are simplified expressions developed in [15] to approximate the correction factor as a function of three parameters: the number of heat transfer units, the number of turns, and the heat capacity rate ratio.

Spiral heat exchangers consist of parallel plates and the space between them is kept by means of bolts that do not have a significant effect upon the thermohydraulic performance of the unit. The geometrical definition of a spiral exchanger requires the specification of: plate spacing of the two streams (δ1 and δ2), plate width (W), inner diameter (Di), outer diameter (Ds), and plate length (L) as shown in **Figure 9**.

The heat transfer performance of SHE is a strong function of the average curvature of the unit represented by the Dean number (K) [16, 17]. As the curvature of a SHE increases, the heat transfer coefficient also does. This effect is more significant in laminar flow regimes than for turbulent regimes [18].

4. Turbulence promoters for tubular geometries

Over the past three decades, a considerable amount of work has been done on the development of turbulent promoters for use in tubular heat exchangers. Since most industrial heat exchangers operate under a turbulent regime, this flow regime has been the focus of the research. The turbulence promoter most widely known and applied in industrial applications is by far the wire matrix type. One of the most successful commercial designs is called HiTRAN [19]. **Figure 10** shows the mechanical construction of this type of insert.

In operation, a wire matrix turbulence promoter makes the most of the velocity profiles of a fluid flowing inside a tube. The center of the tube, where the highest velocity occurs, hits the central part of the insert were the matrix is denser than the sections near the walls. The higher velocity fluid is redirected in the radial direction to the walls where the laminar sublayer is fully removed. The final effect is the increase of the heat transfer coefficient.

Many new types of turbulence promoters have been developed recently. Comment types are: twisted tapes [20–24], winglets tapes [25, 26], circular rings [23, 27], horseshoe baffles [28],

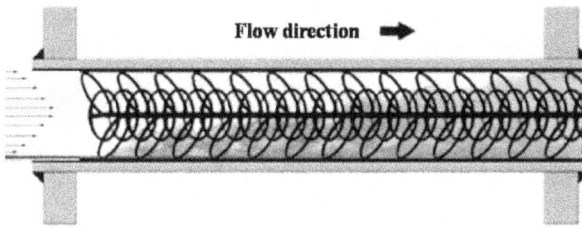

Figure 10. Mechanical features of the wire matrix turbulence promoter (HiTRAN).

helical inserts [29], and coil wires [30]. The optimization of their design for improving thermo-hydraulic performance has been the focus of research. For instance, in [20] modifications that included different size perforations along the tape were made. Another variation was implemented in [21] through the incorporation of pins into the twisted tapes. Other designs such as the proposed in [23] consists of inserts formed by circular-rings combined with twisted tapes. Alternatively, the twisted tapes have been modified including v-cuts along the length [22]. Other varieties of twisted tape technology have been reported in [23, 24].

The modification of basic insert geometries is done seeking to improve on its thermohydraulic performance as is the case of the delta-winglet tape where a considerable augmentation of the heat transfer and friction factor rate has been reported [25, 26]. Some other geometries exhibit improved thermal performance but at the cost of considerable increase of pressure drop. This case was reported in [27] for the v-shaped rings. Other types of geometries are: the inclined horseshoe [28] and the triple helical tape insert [29]. Most thermohydraulic performance expressions derived from experimental work have derived using water and air as thermal fluids under turbulent conditions.

Passive process intensification methods are a very effective tool to reduce the size of new heat exchangers or to enhance the capacity of existing units. The performance of turbulence promoters under turbulent regime is studied based on the coefficient of thermal improvement (η). This parameter provides a means of interpreting the improvement in heat transfer and pressure drop and serves as a performance comparison factor for design and selection purposes. A sample of the many turbulence promoter geometries encountered in the open literature is shown in **Table 1**. The selection includes those geometries that exhibit high thermal performance and low pressure drop. The heat transfer improvement of these systems goes from 30 to 500%. The recommended value of the coefficient of thermal improvement is one or above. The data reported in **Table 1** correspond to those systems with the largest η values which are recommended for industrial applications.

4.1. Performance comparison

The performance parameter used to compare the thermohydraulic performance of turbulence promoters is the thermal enhancement factor or coefficient of thermal enhancement (η). This parameter is determined assuming the same pumping power in the bare tube and the tube with inserts. This is expressed as:

$$\left(\dot{V} \cdot \Delta P\right)_p = \left(\dot{V} \cdot \Delta P\right)_s \tag{7}$$

Turbulence promoter configuration	Reference	Name	Working fluid	Parameters	Reynolds range	Nup/Nus	fp/fs	TEF (η)
a	[20]	Perforated twisted tape	Air, turbulent	Rp = 1.6–14.7%	7200–49,800	2.07–4	2.1–4.6	1.28–1.59
b	[21]	Twisted tape with wire-nails	Air, turbulent	y = 2, 4.4, 6. Lw = 14 mm dw = 2 mm dh = 4 mm	2000–12,000	1.66–1.93	2.67–5.8	1.06–1.30
	[22]	Circular-rings and twisted tapes	Air, turbulent	y/W = 3, 4, 5 l/Di = 1, 1.5, 2	6000–20,000	CR alone 2.36–2.80 CR and TT 2.44–4.70	CR alone 7.4–13.75 CR and TT 11.94–35.83	CR alone 1.09–1.25 CR and TT 1.04–1.42
	[29]	Triple helical tape insert	Air, turbulent	α = 9, 13, 17, 21° rod = 12 mm	22,000–51,000	2.75–4.50	1.9–3.0	2.15–3.70
	[22]	V-cut twisted tape	Water, turbulent	y = 2, 4.4, 6 de/W = 0.34, 0.43 w/W = 0.34, 0.43	2000–20,000	1.36–2.46	2.49–5.82	1.07–1.27

Turbulence promoter configuration	Reference	Name	Working fluid	Parameters	Reynolds range	Nup/Nus	fp/fs	TEF (η)
	[24]	TT with alternate axes with different wings	Water, turbulent	d/W = 0.1, 0.2, 0.3; b/W = 0.2; y/W = 4.0	5500–20,200	Tra; 1.74–2.85; Rec; 1.68–2.64; Tri; 1.62–2.49	Tra; 4.35–7.99; Rec; 3.83–6.72; Tri; 3.54–6.26	Tra; 1.06–1.42; Rec; 1.05–1.39; Tri; 1.04–1.35
	[25]	Staggered-winglet perforated-tapes (WPT) and staggered-winglet tape (WTT)	Air, turbulent	B_R = 0.1, 0.15, 0.2, 0.25, 0.3; P_R = 0.5, 1.0, 1.5; α = 30°; A_h/A_t = 0.125	4180–26,000	WPT; 2.39–4.78; WTT; 4.63–4.90	WPT; 4.87–42.69; WTT; 33.46–49.80	WPT; 1.23–1.71; WTT; 1.26–1.52
	[27]	V-shaped rings	Air, turbulent	RB = 0.1, 0.15, 0.2; Rp = 0.5, 1.0, 1.5, 2.0; α = 30°	5000–25,000	2.47–5.77	6.57–82.01	1.15–1.63
	[28]	Inclined horseshoe baffles	Air, turbulent	B_R = 0.1, 0.15, 0.2; P_R = 0.5, 1.0, 2.0; α = 20°, 45°	5300–24,000	20°; 1.92–3.08; 45°; 2.56–3.10	20°; 1.78–6.76; 45°; 3.16–6.84	20°; 1.34–1.92; 45°; 1.29–1.82

Table 1. Thermo-hydraulic parameters of turbulence promoters reported in the literature.

The relationship between the friction factor and the Reynolds number is given by:

$$(f \cdot Re^3)_p = (f \cdot Re^3)_s \tag{8}$$

Or rearranging by:

$$Re_s = Re_p \left(\frac{f_p}{f_s}\right)^{\frac{1}{3}} \tag{9}$$

The thermal improvement factor (η) is the ratio between the heat transfer coefficient, hp, of the tube with the insert and that of the bare tube, hs.

$$\eta = \left.\frac{h_p}{h_s}\right|_p = \frac{\left(\frac{Nu_p}{Nu_s}\right)}{\left(\frac{f_p}{f_s}\right)^{\frac{1}{3}}} \tag{10}$$

The value η reduces as Re increases, except for the triple helical tape insert where the factor increases with Re.

4.2. Guide to insert selection

The selection of the suitable turbulence promoter for a given application must consider the following criteria:

1. The values of η must be larger than one. Values above this figure, indicate that more heat can be recovered for the same geometry and pumping power.

2. The Reynolds number where the insert operates must be within the same range as the data is available.

3. The operating pressures must be within the limits determined by the materials of construction.

5. Conclusions

The main resistance to heat transfer in conventional heat exchangers is the thermal conduction through the laminar sublayer attached to the surface. Improvement of the heat transfer rate involves the removal of this layer at the expense of increased pressure drop. Heat transfer enhancement techniques can be applied at the design stage of new units or in the retrofit of existing units. In design, fluid velocity is a degree of freedom that can be manipulated by appropriate choice of the exchanger dimensions related to cross sectional area. Alternatively, mechanical devices such as inserts are available to promote local turbulence and increase the heat transfer rate. Such devices can also be used in retrofit for increased heat recovery or increased production. New exchanger technology has emerged to provide alternative solutions to accomplish the following goals: (1) to achieve the given heat load within the limitations imposed by pressure drop in the smaller heat transfer equipment, or (2) increase the heat load within the limitations of pressure drop for the same installed heat transfer area.

New exchanger technology is evolving in the direction of more compact surfaces. A compact surface is designed such that the thermohydraulic performance shows higher heat transfer rate and reduced pressure drop. One of the main problems still to overcome with compact surfaces is the limitations they have in terms of the operating conditions that can withstand, since they cannot operate at high temperatures and pressures. Research and development in this area are focused in the development of new geometries and materials of construction.

Nomenclature

b	plate spacing (mm)
Di	inner diameter (m)
Ds	outer diameter (m)
Dp	port diameter (m)
dh	hydraulic diameter (m)
h	heat transfer coefficient (W/m² °C)
F	ratio of new velocity to original velocity
f	friction factor (–)
fs	ratio of secondary surface to total surface area (–)
K	parameter in correlation for heat transfer coefficient
Kp	parameter in expression for pressure drop
k	fin thermal conductivity (W/m °C)
L	plate length (m)
N	number of thermal plates (–)
Nu	Nusselt number (–)
P	number of passages (–)
Re	Reynolds number (–)
$\dot{V}$	volumetric flow rate (m³/s)
v	fluid velocity (m/s)
W	plate width (m)

Greek letters

β	helical baffle angle, chevron angle (°)
βs	surface area to volume between plates (m^2/m^3)
δ	plate spacing (mm)
η	thermal enhancement factor (–)
φ	area enlargement factor (–)
ΔP	pressure drop (Pa)
τ	fin thickness, plate thickness (mm)

Subscripts

p	bare tube
t	tube side
s	shell side, tube with inserts

Superscripts

N	new condition
0	original condition

Author details

Martín Picón-Núñez*, Jorge C. Melo-González and Jorge Luis García-Castillo

*Address all correspondence to: picon@ugto.mx

Department of Chemical Engineering, University of Guanajuato, Mexico

References

[1] Hesselgreaves JE. Compact Heat Exchanger: Selection, Design and Operation. 1st ed. Oxford: Pergamon; 2001

[2] Polley GT, Reyes Athie CN, Gough M. Use of heat transfer enhancement in process integration. Heat Recovery Systems and CHP. 1992;12(3):191-202

[3] Sivarajan C, Rajasekaran B, Krishnamohan N. Numerical and experimental study of helix heat exchanger. International Journal of Engineering Research and Technology. 2014;2(4):2278-0181

[4]   Movassang SZ, Taher FN, Razmi K, Azar RT. Tube bundle replacement for segmental and helical shell and tube heat exchangers. Applied Thermal Engineering. 2013;**51**:1162-1169

[5]   Zhang L, Xiaa Y, Jianga B, Xiaoa X, Yang X. Pilot experimental study on shell and tube heat exchangers with small-angles helical baffles. Chemical Engineering and Processing. 2013;**69**:112-118

[6]   Chen GD, Zeng M, Wang QW, Qi SZ. Numerical studies on combined parallel multiple shell-pass shell-and-tube heat exchangers with continuous helical baffles. Chemical Engineering Transactions. 2010;**21**:229-234

[7]   Gao B, Bi Q, Nie Z, Wu J. Experimental study of effects of baffle helix angle on shell-side performance and shell-and-tube heat exchangers with discontinuous helical baffles. Experimental Thermal and Fluid Science. 2015;**68**:48-57

[8]   Picón-Núñez M, García-Castillo JL, Alvarado-Briones B. Thermo-hydraulic design of single and multi-pass helical baffle heat exchangers. Applied Thermal Engineering. 2016;**105**(25):783-791

[9]   Kays WM, London AL. Compact Heat Exchangers. 3rd ed. EUA: McGraw Hill; 1984

[10] Marriott J. Recover more heat with plate heat exchangers. Chemical Engineering. 1971; **78**(8):127-134

[11] Picón-Núñez M. Heating and Cooling Systems Analysis Based on Complete Process Network. In: Rahman MS, Ahmed J. Handbook of Food Process Design. 1st editors. Oxford, UK: Black Well Publishing Limited. March 2012. p. 299-334

[12] Ayub ZH. Plate heat exchanger literature survey and new heat transfer and pressure drop correlations for refrigerant evaporators. Heat Transfer Engineering. 2003;**25**(4):3-16

[13] Prabhakara BR, Krishna PK, Sarit KD. Effect of flow distribution to the channels on the thermal performance of a plate heat exchanger. Chemical Engineering and Processing. 2002;**41**:49-58

[14] Gut JAW, Pinto JM. Modeling of plate heat exchangers with generalized configurations. International Journal of Heat and Mass Transfer. 2003;**46**:2571-2585

[15] Wilhelmsson B. Consider spiral heat exchangers for fouling applications. In: Hydrocarbon Processing. Houston, Texas: Gulf Publishing Company; 2005. pp. 81-83

[16] Martin H. Heat Exchangers. 1st ed. Hemisphere Publishing Corporation; New York, USA; 1992. pp. 73-82

[17] Minton PE. Designing spiral heat exchangers. Chemical Engineering. 1970;**77**:103-112

[18] Egner MW, Burmeister LC. Heat transfer for laminar flow in spiral ducts of rectangular cross section. Journal of Heat Transfer. 2005;**27**:352-356

[19] Drögemüller P. The use of HitRAN wire matrix elements to improve the thermal efficiency of tubular heat exchangers in single and two-phase flow. Chemie-Ingenieur-Technik. 2015;**87**(3):188-202

[20] Bhuiya MMK, Chowdhury MSU, Saha M. Heat transfer and friction factor characteristics in turbulent flow through a tube fitted with perforated twisted tape inserts. International Communications in Heat and Mass Transfer. 2013;**46**:49-57

[21] Murugesan P, Mayilsamy K, Suresh S. Heat transfer and friction factor studies in a circular tube fitted with twisted tape consisting of wire-nails. Chinese Journal of Chemical Engineering. 2010;**18**:1038-1042

[22] Murugesan P, Mayilsamy K, Suresh S, Srinivasan PSS. Heat transfer and pressure drop characteristics in a circular tube fitted with and without V-cut twisted tape insert. International Communications in Heat and Mass Transfer. 2011;**38**:329-334

[23] Eiamsa-ard S, Kongkaitpaiboon V, Nanan K. Thermohydraulics of turbulent flow through heat exchanger tubes fitted with circular-rings and twisted tapes. Chinese Journal of Chemical Engineering. 2013;**21**:585-593

[24] Wongcharee K, Eiamsa-ard S. Heat transfer enhancement by twisted tapes with alternate-axes and triangular, rectangular and trapezoidal wings. Chemical Engineering and Processing Process Intensification. 2011;**50**:211-219

[25] Skullong S, Promvonge P, Thianpong C, Pimsarn M. Heat transfer and turbulent flow friction in a round tube with staggered-winglet perforated-tapes. International Journal of Heat and Mass Transfer. 2016;**95**:230-242

[26] Skullong S, Promvonge P, Thianpong C, Jayranaiwachira N. Thermal behaviors in a round tube equipped with quadruple perforated-delta-winglet pairs. Applied Thermal Engineering. 2017;**115**:229-243

[27] Chingtuaythong W, Promvonge P, Thianpong C, Pimsarn M. Heat transfer characterization in a tubular heat exchanger with V-shaped rings. Applied Thermal Engineering. 2017;**110**:1164-1171

[28] Promvonge P, Tamna S, Pimsarn M, Thianpong C. Thermal characterization in a circular tube fitted with inclined horseshoe baffles. Applied Thermal Engineering. 2015;**75**: 1147-1155

[29] Bhuiya MMK, Ahamed JU, Chowdhury SU, Kalam MA. Heat transfer enhancement and development of correlation for turbulent flow through a tube with triple helical tape inserts. International Communications in Heat and Mass Transfer. 2012;**39**:94-101

[30] Gunes S, Ozceyhan V, Buyukalaca O. Heat transfer enhancement in a tube with equilateral triangle cross sectioned coiled wire inserts. Experimental Thermal and Fluid Science. 2010;**34**:684-691

Potentials and Challenges of Additive Manufacturing Technologies for Heat Exchanger

Uwe Scheithauer, Richard Kordaß, Kevin Noack,
Martin F. Eichenauer, Mathias Hartmann,
Johannes Abel, Gregor Ganzer and Daniel Lordick

Additional information is available at the end of the chapter

http://dx.doi.org/10.5772/intechopen.80010

Abstract

The rapid development of additive manufacturing (AM) technologies enables a radical paradigm shift in the construction of heat exchangers. In place of a layout limited to the use of planar or tubular starting materials, heat exchangers can now be optimized, reflecting their function and application in a particular environment. The complexity of form is no longer a restriction but a quality. Instead of brazing elements, resulting in rather inflexible standard components prone to leakages, with AM, we finally can create seamless integrated and custom solutions from monolithic material. To address AM for heat exchangers we both focus on the processes, materials, and connections as well as on the construction abilities within certain modeling and simulation tools. AM is not the total loss of restrictions. Depending on the processes used, delicate constraints have to be considered. But on the other hand, we can access materials, which can operate in a much wider heat range. It is evident that conventional modeling techniques cannot match the requirements of a flexible and adaptive form finding. Instead, we exploit biomimetic and mathematical approaches with parametric modeling. This results in unseen configurations and pushes the limits of how we should think about heat exchangers today.

Keywords: additive manufacturing, computer-aided design, flow simulation, metals, ceramics, fractal geometry

1. Introduction

Heat exchangers are used to transfer heat energy from one to another medium without intermixing them. There are different types available like plate, bundled tube or rotary heat exchangers. **Figure 1** shows an example of a conventional plate heat exchanger.

In addition, heat exchangers also differ in their working principle (counterflow, direct flow, or cross flow) and can consist of differently shaped plates or tubes with, for example, smooth, buckled, or rippled surfaces. A typical wall thickness reaches from 0.4 to 2.5 mm and is mainly designed to withstand blockage, corrosion, active pressure, or abrasive media. Such heat exchangers are very cost-effective.

In conventional heat exchangers, a lot of restrictions and disadvantages exist, concerning the realizable geometry, the operating temperature, as well as the manufacturing costs:

1. Heat exchanger manufactured by the combination of different planar parts is limited in the realizable design and compactness (the ratio between heat exchanging surface and total volume).

2. The assembly of the different parts can result in assembly failures.

3. The realization of mechanical and fluidic interfaces is very challenging (often, the cross-flow principle is realized instead of the superiorly counterflow principle because of the feeding system for the different fluid channels).

4. The joining of the different parts is often realized by brazing. But the brazing material limits the operating temperature, and the brazing process can result in leakage.

5. Because of the used standardized geometries for the parts as well as the whole heat exchanger components, their outer geometry can hardly be individualized. Furthermore, no adjustment of the outer geometry on the shape of the surrounding system can be realized.

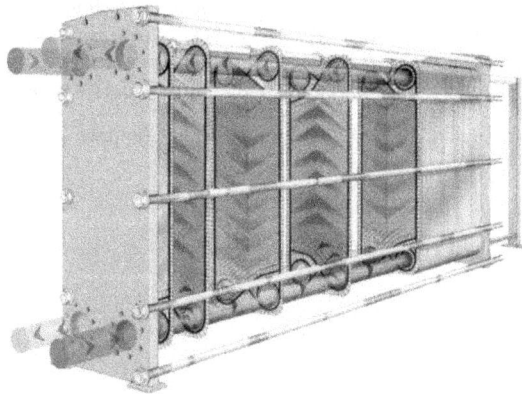

Figure 1. Setup of a conventional plate heat exchanger [1].

6. Some examples for ceramic-based heat exchangers for high temperature or high corrosive or abrasion applications exist, but their design is limited because of the ceramic shaping and finishing technologies. Furthermore, the operation temperature is limited because of the needed joining additives (e.g., solders or brazes) for the different ceramic parts.

Additive manufacturing (AM) is a new class of manufacturing technologies, which has been developed for polymers, metals, and ceramics during the last three decades and keeps evolving. Based on computer-aided design (CAD) files in 3D, typically a layer-wise manufacturing process follows, which allows the realization of component designs as well as inner and outer geometries which were previously regarded as not producible. Concerning the manufacturing of heat exchanger, AM technologies open the door to overcome all of the restrictions mentioned above:

1. The manufacturing of the heat exchanger as one component with integrated mechanical and fluidic interfaces becomes possible.

2. No joining steps are needed, and the same properties are available in the whole component.

3. Very complex designs can be realized, and the ratio between the heat exchanging surfaces to the total volume of the heat exchanger can be increased significantly. The increased performance allows the miniaturization of the heat exchanger.

4. The adjustment of the outer geometry becomes possible, and the required volume for the implementation of heat exchanger and the surrounding system can be decreased.

5. AM of ceramics opens the door for complex heat exchangers for demanding applications concerning operation temperature, abrasion, or corrosion.

Thus, heat exchangers seem to be very interesting for AM while prices are falling especially for AM of metal components [3]. Also, an integrated manufacturing process for heat exchangers seems to be positive on their pressure resistance and against leakage. Today, only a few designs are described or commercially available. EOS and 3TRPD have designed a heat exchanger, which was manufactured with laser beam melting (**Figure 2**). Unfortunately, they have not published any performance data for comparison.

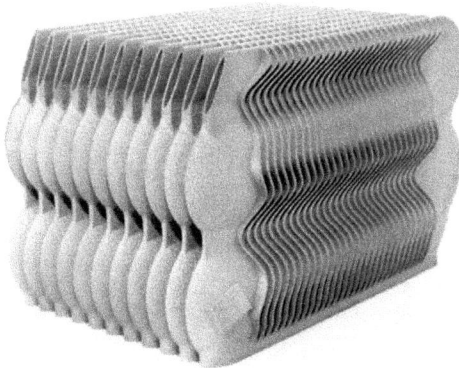

Figure 2. Additively manufactured heat exchanger by EOS and 3TRPD [2].

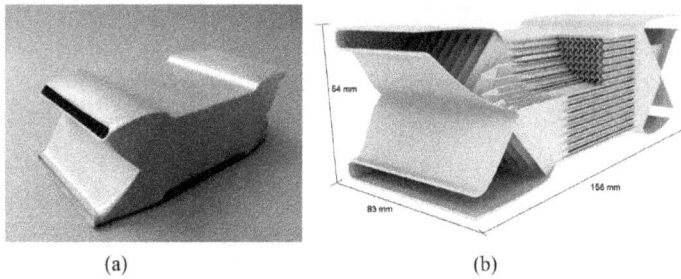

(a)   (b)

Figure 3. Counterflow heat exchanger manufactured by LBM; left: manufactured component [4]; right: rendering of complex internal structure inside the heat exchanger [4].

Furthermore, at Fraunhofer IFAM, a counterflow heat exchanger was developed to improve the efficiency of a micro-gas turbine system (**Figure 3**). In this case study, the hot exhaust gas should heat up the inflowing cold air to improve the overall combustion efficiency of the system. The heat exchanger was particularly designed for laser beam melting, so there was no conventional way to manufacture the part. The heat exchanger combines 18 layers of channels in the limited design space. Furthermore, the complex inner channels were designed in a wave shape combined with a very small spacing to each other in order to maximize the surface for heat transfer (**Figure 3**). With the special design for laser beam melting, it was possible to reduce the required time and costs for postprocessing. It was only necessary to machine the inlet and outlet at the side toward the build platform after the LBM process [C4].

But still, a lot of challenges exist, which have to be overcome, to allow the AM of high-performance heat exchanger.

Design stage:

1. designing and dimensioning of the heat exchanger;

2. generation of the CAD files;

3. modeling and simulation of the fluid flow and heat flux;

4. providing software tools for the different tasks, coupling, and automatization of all tasks.

Manufacturing stage:

1. enhancement of the material portfolios for the different AM technologies;

2. increasing productivity (higher building speed, larger building space with more components manufactured simultaneously, less reject rate) will result in decreasing manufacturing costs.

To overcome the current restrictions, we are working on all links of the process chain. In this chapter, we want to introduce different AM technologies for metals (Laser beam Melting— LBM) and ceramics (Fused Filament Fabrication—FFF and Lithography-based Ceramic Manufacturing—LCM) as well as two different approaches for designing and creating the

CAD files, one based on conventional software tools and one based on mathematical algo-rithms. In addition, the specification of the operation conditions of a solid oxide fuel cell system with an operation temperature of 850°C and higher will exemplarily illustrate the requirements for heat exchangers suitable for high-temperature applications and will justify the need for AM of high-temperature materials like ceramics.

2. Design stage

2.1. Different heat exchanger designs based on conventional CAD tools

For additively manufactured heat exchangers, a high heat exchange capacity is essential. It is represented by the heat flow, which can be calculated from the thermal conductivity λ, the heat transfer area A, the wall thickness s, and the temperature difference ΔT. For different materials, λ is summarized in **Table 1**. It can be seen that the thermal conductivity is an important factor, because it has a linear influence on the heat flow. It becomes evident that steel should not be used as a material for heat exchangers. With copper or silver instead of steel, the heat exchange rate can almost be increased by the factor of 20. But aluminum would be an adequate material of choice, as well.

Usually, the temperature gradient ΔT cannot be influenced by outer parameters and is defined by the process in which the heat exchanger is used. Therefore, the quotient of the heat exchanging area A and the thickness of the wall s are the factors to vary in terms of geometric adaption. The goal for improved structures for a high heat exchange rate should be that they have a high ratio A/s and have a considerable high thermal conductivity. The favorable material becomes aluminum because it can be processed by LBM, and the minimal reproduc-ible wall thickness accounts to 300 μm. Furthermore, the structures have to be self-supporting. Inside the channels, a mechanic finishing process is not feasible, so the wall roughness has to be in an acceptable range, and minimal angles of areas reclining from the normal of the build plate should be lower than 45°.

In this work, principle insights into the design of structures with a high heat exchange capa-bility and coincident low pressure drop shall be given. Therefore, simulation of the flow is inevitable, and the structures were designed using CAD tools. The aim is to obtain a structure,

Material	Thermal conductivity $[\frac{W}{mK}]$
Silver	429
Copper	401
Aluminum	237
High alloyed steel	~20
Low alloyed steel	~30

Table 1. Specific thermal conductivity for different materials [5, 6].

which can be individualized at its outer geometry and optimizes the flow problems inside of the structure.

On the basis of the constraints explained before, basic sketches were developed and compared on its contour length of all the channels that will be involved in the heat transfer. To enhance the performance of the heat exchanger, the aim is to maximize the surface for the heat transfer. Therefore, the contour length of the channels is maximized in each evolution of the sketches by retaining the hydraulic diameter. Basic sketches were made and afterwards extruded in height with a helical-shaped structure for generating a longer streamline and for maximizing the heat transfer area. All in all, seven basic sketches were defined with nine resulting structures, as shown in **Figure 4**. Structures 1 and 2 show a simple geometry, which could also be produced with conventional manufacturing processes. Structures 3–5 are very complicated in terms of connecting geometries. Structures 6.1 and 6.2 are highly optimized for production with straight walls which can be efficiently produced with LBM with special slicer options. At last, structures 7.1 and 7.2 are optimized for production and fluid flow by using special slicer options and not having sharp edges like structures 6.1 and 6.2 which lead to a higher pressure drop.

A comparison of the geometrical and heat flow characteristics of these structures is given in **Table 2**. Also, the structures show a surface roughness as build. This roughness is assumed to be the same value for all side surfaces since they are all in the same build direction (same as the orientation of the pictures).

As stated above, connecting structures for fluid guidance inside the heat exchanging structure are important, but rather complex and difficult to design. As a first attempt for structures 5 and 6.2, connecting structures are shown in **Figure 5**, which were derived from biomimetic role models like fennel.

To avoid such a complexity of the connecting structures and at the same time to enable the engineer to model them economically, a new approach was chosen. The design process now starts with the connecting structures, and the inner geometry is optimized afterwards. The design ideas were adopted from nature, too. Some possible basic structures are shown in **Figure 6**. The purpose of these structures is to transform a round connector with a 6-mm

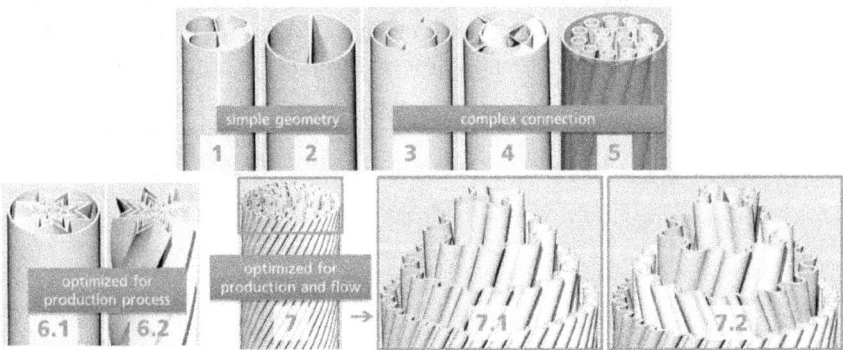

Figure 4. Different designs of inner structures for heat transfer.

	Compactness $[\frac{m2}{m3}]$	Heat flow [W]	Performance per volume $[\frac{W}{cm3}]$	Pressure drop [Pa]
1	267	270	38.20	0.05
2	243	170	24.05	0.06
3	664	200	28.29	0.06
4	651	350	49.51	0.50
5	1576	420	59.42	0.15
6.1	1471	520	65.47	0.10
6.2	**1711**	550	**80.54**	0.20
7.1	1264	750	50.71	**0.10**
7.2	1246	**900**	60.86	**0.10**

Table 2. Simulation results for the different designed structures (compactness, heat flow, performance per volume and pressure drop; possible choice highlighted).

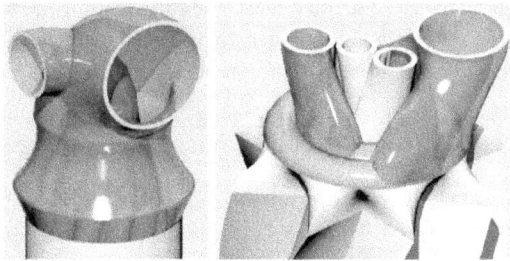

Figure 5. Exemplary connecting structures for designed structures 5 and 6.2.

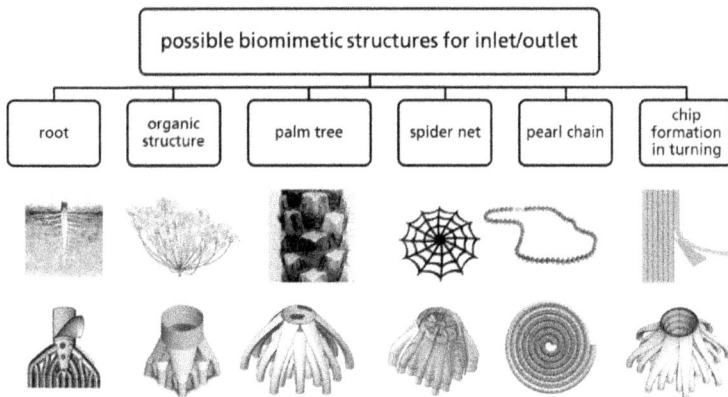

Figure 6. Sampled possible structures for connecting geometries adopted from nature.

diameter to any amount of complex-shaped inner structures like tubes. In addition, the flow has to be equally distributed in all inner tubes, and heat transfer should also start within the connecting structure to minimize losses. Certainly, manufacturability has to be guaranteed as well by allowing a minimum angle of 45° to the building plane. For the same reason, gaps inside the structures have to be avoided, and the connecting structures have to be as narrow as possible.

It is obvious that from these connecting structures, such immersed structures like those shown in **Figure 4** (structures 6 and 7) cannot be accessed. This means that ideally nestable inner structures are preferred, which can be enveloped with one larger outer channel to gain a tube-in-tube-like design known from conventional heat exchangers.

Designing the inner structures, which are applicable to the previous connecting structures, is the next step. The profile shape may not be too complex due to terms of a steady connection to connecting structures. Furthermore, analytical calculation cannot be used for dimensioning these structures since they should all have the same hydraulic diameter. Therefore, fluid simulation was used for dimensioning. To validate these findings, experimental parameter evaluation should be conducted in further investigations. In **Figure 7**, some possible designed inner structures are compared. Herein, the lowest pressure drop per performance as indicating value was used to choose the optimal inner structure. These profiles are based on mathematical algorithms after Sierpinski (left profile) and Hilpert (right profile).

Furthermore, the heat exchanging volume has to be filled with the inner structures using optimal arrangement options to fill the volume using curves, planes, or lines.

For combining the presented structures to a whole heat exchanger, a connecting structure has to be chosen and specified. In **Figure 8**, the possible structures emerged from biomimetic structures of **Figure 3** are preselected in terms of producibility, compactness, and capability. The final selection was done by simulating the connecting structures with CFD using Ansys CFX to gain a uniform flow distribution over all profiles and minimum losses in flow distribution.

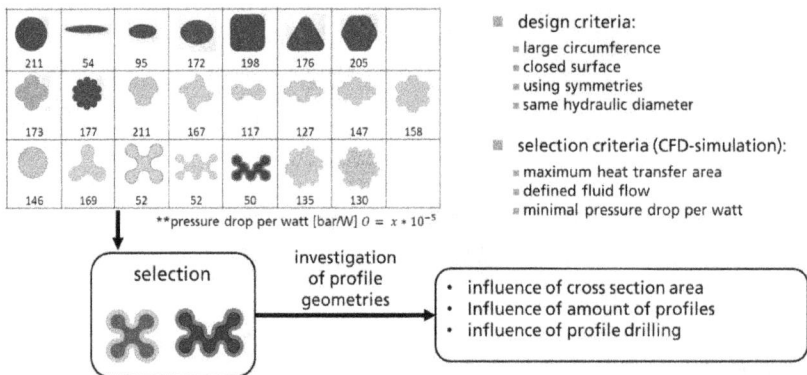

Figure 7. Comparison between inner structures with the same hydraulic diameter and optimal designs selected using fluid simulation.

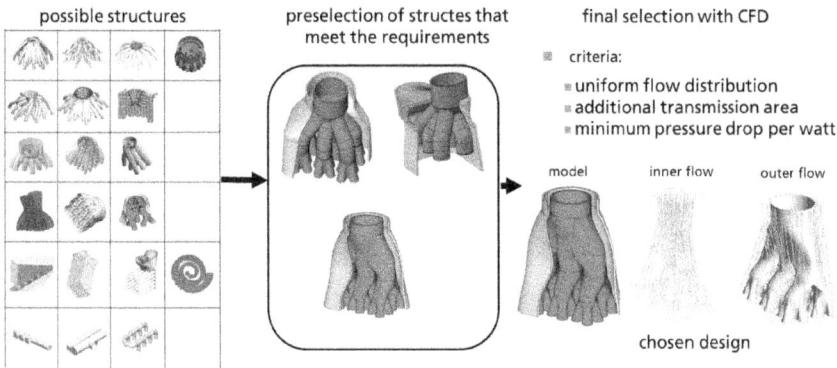

Figure 8. Selection of an optimal connecting structure.

Finally, three different complete heat exchanging structures were designed for comparison and to gain an optimal structure. The structures were combined to a whole exchanger geometry, and the capability of the models was simulated as shown in **Figure 9**.

As depicted, a uniform flow distribution can be achieved in the inner flow. The outer flow, however, shows a highly turbulent and uneven distribution especially in designs 2 and 3, leading to a high pressure drop. The obtained values from these structures are depicted in **Table 3**. The different values for design no. 1 are based on different lengths of the inner structure (50, 100, and 150 mm).

It can be seen that the compactness is highly dependent on the length of the structure (design no. 1) because the influence of connecting structures decreases with an increasing length. With this example, a good scalability with different performances can be achieved, especially in terms of individualization. All in all, individualized and complex-shaped heat exchanging

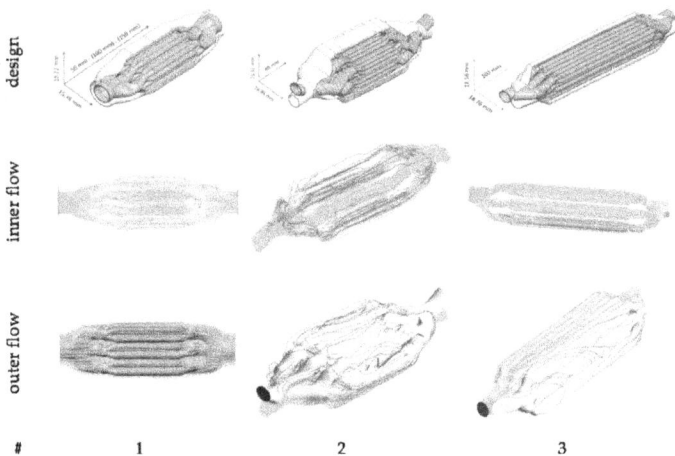

Figure 9. Whole heat exchanger structures and flow in inner and outer channels.

	1	2	3
Compactness [m^2/m^3]	736–1291–**1348**	735	1161
Heat flow $\dot{Q}$ [W]	3465–6202–**9372**	4482	6985
Capacity per volume $\dot{Q}/V$ [W/cm^3]	376–**435**–421	139	272
Pressure drop Δp [bar]	1.87–1.95–2.50	**5.49**	**0.97** (inner), **2.58** (outer)

Table 3. Calculated values for designed heat exchangers (outstanding values highlighted).

structures can be obtained for optimized production with additive manufacturing. But still, optimization has to be done to increase the performance and to lower the pressure drop. Also, experimental validation of the simulation has to be carried out even since the convergence in simulation was not satisfying.

2.2. Alternative ways to generate new designs for heat exchanger

2.2.1. New approach

In the following section, a new approach to generate efficient design structures for heat exchangers is presented. This new approach is one of the main topics of the instaf project.

Fractal macrostructures are used to generate a large inner surface, which implicates a better energy transfer between the heat-exchanging fluids. The creation of microstructures with roughness (with respect to the process-based roughness of AM), induced by partial Brownian motions, leads to turbulences. This raises the performance even more.

Two irreconcilable goals define the design scope. On the one hand, the surface for heat exchanging should be maximal and the fluids should remain in the heat exchanger as long as possible. But on the other hand, the restrictions concerning the manufacturing process (e.g., minimal wall thickness, resolution, waiver of support structures, etc.) and the operation as heat exchanger (e.g., pressure drop, mechanical strength, etc.) have to be considered as well.

2.2.2. Space-filling curves

To generate a maximum heat exchange surface, the so-called fractals were studied. Fractals are inspired by nature and are a branch of research essentially introduced to mathematics by Benoit Mandelbrot [7]. Stochastic fractals can be found in lung alveoli and other breathing organs or in the capillaries in the fin of whales with a heat- and energy-saving component.

For the design of the macrostructures, our special focus lies on the construction of space-filling curves, which belong to the group of FASS curves. The acronym FASS stands for space-filling, self-avoiding, simple, and self-similar. This class of curves traverses every vertex of a polygonal grid so that every point is reached once. Because they are not allowed to cut themselves, they separate to areas perfectly and are therefore well suited for the construction of heat exchangers.

A Lindenmayer system (L-system) was used to generate the curves. An L-system is a way to describe a repetitive structure with a small number of rules. It is a character-based rewriting system, which consists of constants. These are representing draw commands and variables, which are replaced in every iteration step through a replacement rule. In **Figure 10**, the system is visualized by means of the Hilbert curve with the variables X, Y, the constants F (straight line), −(clockwise rotation), and +(counterclockwise rotation) with a starting value of X. The rules are X → +YF − XFX − FY+ and Y → −XF + YFY + FX−.

A process was developed to generate a large number of different curves with a small number of basic motifs considering only 2 × 2- and 3 × 3 grids in order to investigate the best properties. There are four ways to traverse through these grids, which leads to seven basic motifs, paying attention to reflection. Every motif can also be used as a mapping structure. Therefore, it represents the connecting pieces of the following iteration step. **Figure 11** shows a curve where the basic motif of a Peano curve (represented by the purple lines) and the mapping structure of the Hilbert curve (represented by the green lines) are combined. The

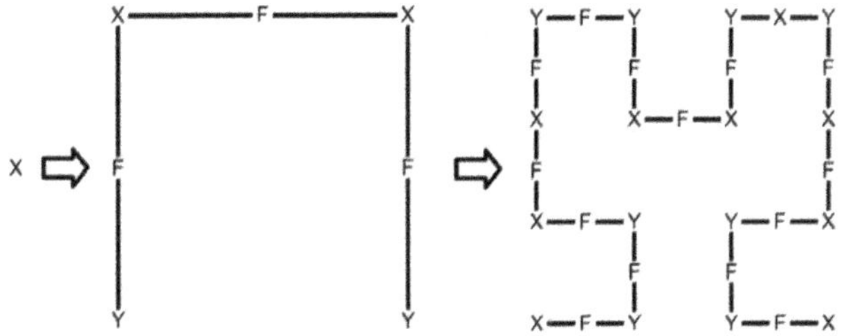

Figure 10. A visualization of the Lindenmayer-code demonstrated by means of the Hilbert Curve.

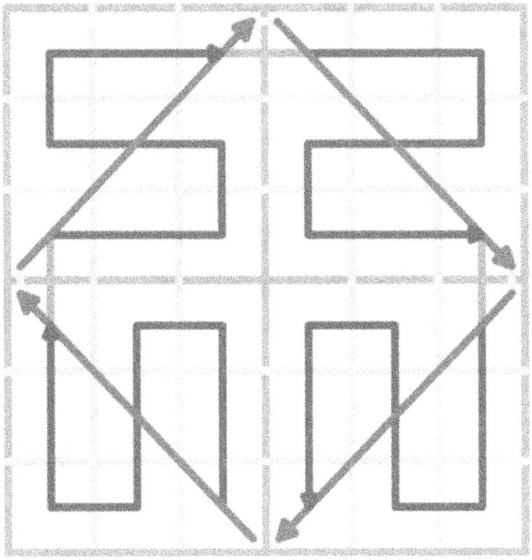

Figure 11. One possible combination of the motifs.

ratio of curve length to the surface area or in higher dimension from surface to volume is always the same for the same grid size. Therefore, the criteria of evaluation are turbulences in flow and the velocity of the fluids so as the basic conditions of the used AM technology.

2.2.3. Generation of CAD files

To construct a three-dimensional structure, successive iteration steps of a curve were placed in a predefined distance. They were combined with an NURBS-based (non-uniform rational basic spline) surface. The surface is lofted over the curves and defines a closed structure in combination with the outer skin. **Figure 12** illustrates the process for the first four iteration steps of a Peano curve. Since separate areas should always be consistent, the offset must be adjusted. As the curve gets longer in each iteration step, the selected offset also decreases. **Figure 13** demonstrates the result, visualizing one of the two separate liquids.

In the illustrated structure, the thickness of the partition wall decreases from 4.5 to 0.2 mm, which corresponds to a factor of 22.5. The length of the partition wall increases in the same range, simultaneously. In addition, the extrusion of the final geometry results in a heat exchanger with a compactness of about 3000 m^2/m^3, which can be operated as a counterflow heat exchanger. The structure was created using the CAD software Rhinoceros 3D. In **Figure 13**, some renderings of the fluid paths are shown.

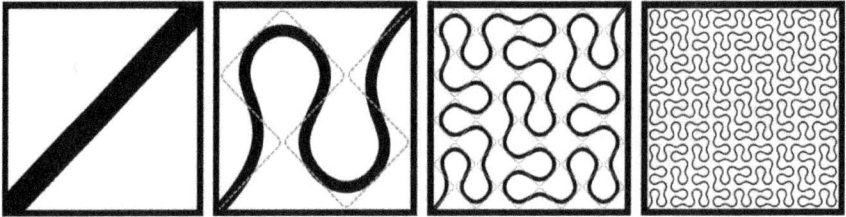

Figure 12. The first four iteration steps of the Peano curve. They lead to a feeding structure for which we combined the four layers with an NURBS-based surface.

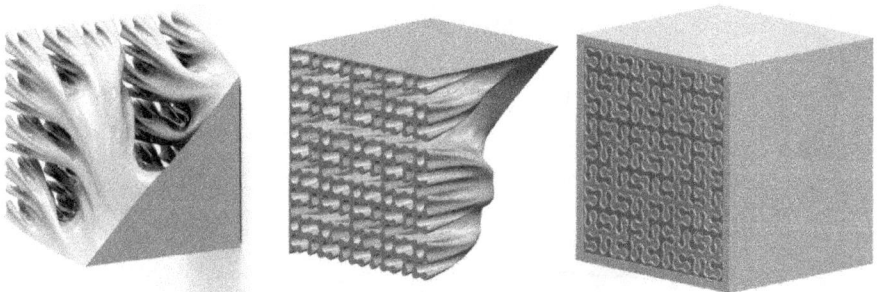

Figure 13. Rendering of the dispersion of one fluid in a feeding structure designed with a Peano curve down to the fourth iteration step.

Figure 14. Alternative structure as a concept for a heat exchanger. From left to right: a 5/8 cutout to show the inner structure; the whole component (lying); the curves used for designing the structure.

Another structure for a heat exchanger is presented in **Figure 14**. In this case, a curve was chosen, which can be closed. It can be used without an outer wrapping, and this is why it could also be used as an immersion heater. For a better view into the internal structure, it is presented cut open. The outer structure could be represented by a cuboid. The curves are constructed with the end points of the elements as control points. This makes the curves even longer and smoothens them evenly.

3. Laser beam melting (LBM): AM of metal components

The LBM process has a lot of different specific names such as LaserCUSING®, selective laser melting (SLM®), direct metal laser sintering (DMLS®), and direct metal printing, to name only a few. All of these names describe the powder bed-based laser process, where a part is manufactured by means of thin layers of a powder material, which are applied by a scraper and molten selectively by laser energy.

The digital process chain begins with the 3D CAD file (*.stl) of the part, which has to be manufactured. This file is transferred to a software program where the support generation and the positioning in the building chamber of the machine are done. Afterwards, the so-called build job is sliced into layers of 20–100 μm, dependent on the material and the laser parameters and carried over to the machine [8]. A principle schematic of an LBM machine is shown in **Figure 15**.

To avoid oxidation, the process itself as well as the preparation and postprocessing of the powder has to be done under inert gas atmosphere. Overhanging structures have to be supported by supporting structures which have to be produced as well for stabilizing the model and to improve the dissipation of heat below these geometries [9].

The process is suitable for producing individual and highly complex parts and hollow structures such as topology-optimized components as shown in **Figure 16**. Also, very fine structures like lattice structures can be produced, which is also depicted therein.

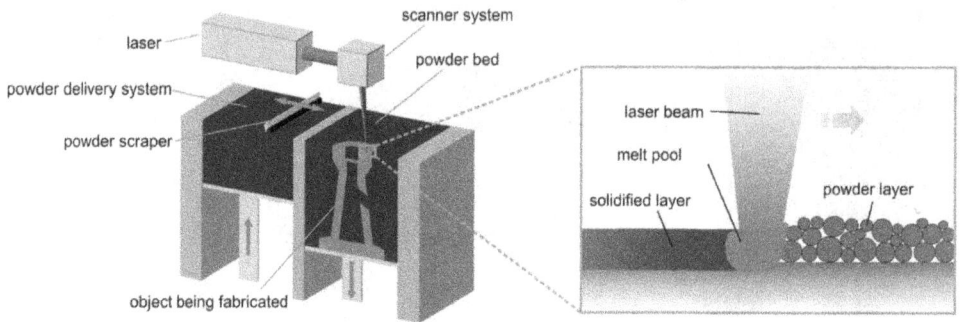

Figure 15. A principle schematic of an LBM machine.

Figure 16. Complexly designed skateboard trunk manufactured with LBM.

After the manufacturing process, the part is separated from the build platform, and the support structure is removed. To adjust the mechanical properties of the part, a heat treatment can be applied. Other conventional methods like machining or polishing can be utilized to achieve a better surface quality, dependent on the requirements.

Depending on slicing and scanning strategy, the quality of manufactured parts can widely vary, concerning cracks, pores, residual stresses, distortions, tightness, and fatigue properties. However, LBM processes are becoming relevant in series and tool production, while the reliability of such manufacturing technologies and the resulting component quality are of high importance [10].

At Fraunhofer IWU, a counterflow heat exchanger was specially developed for the LBM process with the focus on a low pressure drop, flexible, as well as compact design (**Figure 17**). To reduce the pressure drop and validate the best version, a fluid analysis was executed on each design. To get a maximum heat transfer and integrated insulation to the outer atmosphere, the cold channel is wrapped around the hot channel within the heat exchanger (heat exchange surface/volume: 405 m^2/m^3; performance: 36.2 W/cm^3). Also, the wall between the cold and the hot channel system is reduced to a minimum of 0.8 mm to enhance the heat transfer. As a result of the special design for the laser beam melting process, no support

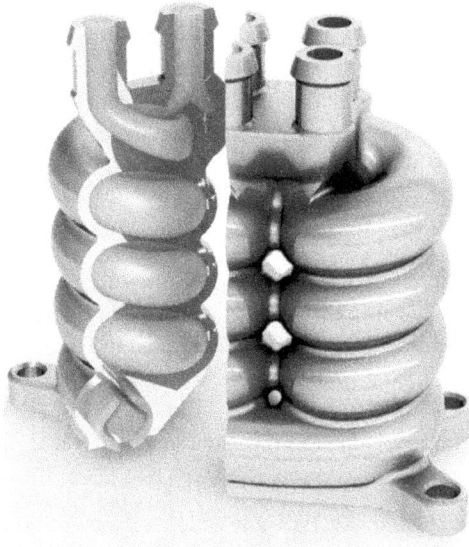

Figure 17. A counterflow heat exchanger with a low pressure drop and compact, flexible design.

structure and postprocessing were needed to manufacture the heat exchanger. The pressure drop (0.20 bar) as well as heat flow (1.3 kW) of the final design was calculated with flow simulations using Ansys CFX. The propagated heat exchanger shows one imperfection since it produces a thermal short circuit in the area of the connection geometries.

The next step will be to avoid the thermal short circuit and to implement the designs described before within the metal heat exchanger in consideration of the general conditions of LBM.

4. AM of ceramic components

4.1. Operating conditions in a high-temperature fuel cell system

The research and development of new technologies is sometimes limited by the availability of commercial peripheral components. These new technologies change the requirements on state-of-the-art products.

One of those new technologies is the fuel cell. A fuel cell has the benefit of directly converting chemical energy to electrical power without the conventional process steps in between. This reduces the losses on the whole conversion path. Fuel cells are available in various types with different characteristics (**Figure 5**). The low-temperature fuel cells are mostly used for portable or mobile applications because of the high power density, for example, as a battery charger for mobile devices. High-temperature fuel cells are typically used for stationary applications like power and heat supply for households and facilities [11].

Most of the low-temperature fuel cells like the PEFC (polymer electrolyte fuel cell) need pure hydrogen or at least "clean" fuel. Not allowed substances like carbon monoxide have to be filtered before operation. The high-temperature solid oxide fuel cell (SOFC) is designed to generate electrical and/or thermal power with a high efficiency based on the usage of world-wide available fuels such as natural gas and LPG (liquefied petroleum gas). These fuels are reformed to a gas mixture of hydrogen and carbon monoxide directly inside the SOFC System. In contrast to most of the common fuel cells, an SOFC can also generate electrical power by carbon monoxide conversion (**Figure 18**).

In order to achieve this high efficiency, high temperatures are necessary. The standard operation temperature of an SOFC is above 700°C. As mentioned before, these systems have different gas processing steps included [12], which is illustrated in **Figure 19** as well:

1. the reforming process (REF),

2. the conversion of hydrogen and carbon monoxide to water and carbon dioxide inside the fuel cell itself, and

3. furthermore, a postprocessing step typically called tail gas oxidation (TOX).

The fuel cell itself cannot convert 100% of the fuel for various reasons, which will not be further explained in this context but are discussed in [13]. The rest of the fuel has to be oxidized in order to avoid emission of hydrogen and carbon monoxide. This kind of reaction leads to temperatures of 900°C and higher.

The different reactors REF and SOFC have different requirements on heat treatment. The REF needs heat for a high efficiency. The SOFC receives the reaction products of the REF with a temperature of approx. 800°C on the anode side. On the cathode side, typically air is used. In order to avoid thermal stress and to realize the necessary operating temperature of the SOFC, this air has to be preheated.

		Single chemical reaction (Anode / Cathode)	Charge carrier	Reformate components allowed	not allowed	Operating temperature
PEFC	A	$H_2 \rightarrow 2H^+ + 2e^-$	$\downarrow H^+$	CO_2	CH_4, CO, S	80 °C
	C	$\frac{1}{2}O_2 + 2H^+ + 2e^- \rightarrow H_2O$				
DMFC	A	$CH_3OH + H_2O \rightarrow CO_2 + 6H^+ + 6e^-$	$\downarrow H^+$	(CO_2)	CO, S	60..130 °C
	C	$\frac{1}{2}O_2 + 6H^+ + 6e^- \rightarrow 3H_2O$				
PAFC	A	$H_2 \rightarrow 2H^+ + 2e^-$	$\downarrow H^+$	CH_4, CO_2	CO, S	200 °C
	C	$\frac{1}{2}O_2 + 2H^+ + 2e^- \rightarrow H_2O$				
AFC	A	$H_2 + 2(OH)^- \rightarrow 2H_2O + 2e^-$	$\uparrow OH^-$	-	CH_4, CO_2, CO, S	65..220 °C
	C	$\frac{1}{2}O_2 + H_2O + 2e^- \rightarrow 2(OH)^-$				
MCFC	A	$H_2 + CO_3^{2-} \rightarrow H_2O + CO_2 + 2e^-$ $CO + CO_3^{2-} \rightarrow 2CO_2 + 2e^-$	$\uparrow CO_3^{2-}$	CH_4, CO_2, CO	S (>0,5 ppm)	650 °C
	C	$O_2 + 2CO_2 + 4e^- \rightarrow 2CO_3^{2-}$				
SOFC	A	$H_2 + O^{2-} \rightarrow H_2O + 2e^-$ $CO + O^{2-} \rightarrow CO_2 + 2e^-$	$\uparrow O^{2-}$	CH_4, CO_2, CO	S (>1,0 ppm)	700..1000 °C
	C	$O_2 + 4e^- \rightarrow 2O^{2-}$				

Figure 18. An overview of fuel cell technologies, including temperatures, reactions, and allowed and not allowed chemical components.

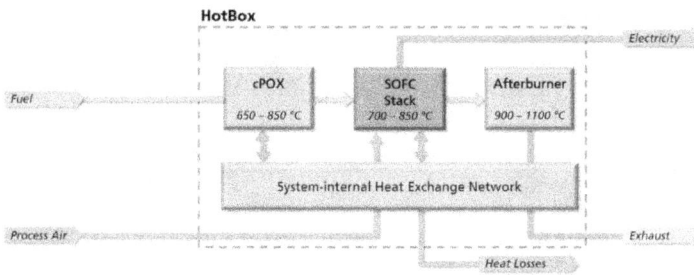

Figure 19. An overview of SOFC system with integrated components.

The realization of the heat treatment requires a rather complex packaging of all components. This packaging is realized in the HotBox. The efficiency of all steps is based on a minimum of losses inside the system. Therefore, an adapted heat exchange from TOX to air and an overall packaging for optimal heat management are necessary.

Commercial heat exchangers are not aware of these requirements. The used materials and/or joining technologies cannot handle this high temperature, and due to the standard design on commercial heat exchangers, especially plate heat exchangers, the integration of such components inside the HotBox is complicated and space-consuming.

In order to achieve the highest possible efficiency of such systems, heat exchangers that combine an adaptable design with high-temperature resistance are indispensable.

4.2. Fused filament fabrication (FFF)

Fused filament fabrication (FFF) is a thermoplastic AM technology which bases on nearly endless filaments which are used as a semi-finished products and which are melted and deposited under a heated nozzle. To generate ceramic components, particle-filled filaments are used to manufacture the so-called green bodies additively [14]. These green bodies have to be debinded, to remove all organic materials, and sintered to densify the microstructure and to achieve the typical ceramic properties. The benefits of this AM technology are the high productivity and the large building space of the available devices. The existing challenges of FFF of ceramic components are the development of highly particle-filled filaments and the defect-free debinding and sintering of the components [14].

SiC filaments were developed to allow the AM of large volume SiC heat exchanger, which can be used for operation temperatures of 1000°C and higher. **Figure 20** shows some ceramic test components and demonstrators manufactured by FFF.

The next steps will be the investigation of FFF with SiC filaments concerning the realizable geometries and the tightness of the sintered structures.

4.3. Lithography-based ceramic manufacturing (LCM)

The LCM technology was developed and commercialized by Lithoz GmbH, Austria [15]. As a special kind of stereolithography, free radical polymerization of the binder system takes place

Figure 20. Different ceramic demonstrators additively manufactured by FFF; left: SiC (green state); right: various ceramics (sintered state).

with light of a defined wavelength, causing the suspension to solidify. Via a DLP module, the suspension is selectively irradiated with a blue light, whereby all areas to be cross-linked on a given plane are exposed at the same time. The ceramic particles dispersed in the suspensions are fixed in the solid polymer matrix (green body). A final debinding and sintering step is necessary for this AM technologies for ceramics, as well [16].

The LCM technology impresses with a very high resolution (wall thickness down to 100 μm possible) [16] and very good surface properties (R_a < 1 μm) of the sintered components. The challenges are the small building area (($76 \times 43 \times 150$) mm^3) and the low productivity, both resulting in relatively high manufacturing costs as well as the cleaning and the debinding process for the green components [17].

Both FASS structures which were described before were additively manufactured via LCM technology and Al$_2$O$_3$ suspension of Lithoz. The sintering occurred at 1650°C which allows operation temperatures of significantly more than 1000°C. **Figure 21** shows the feeding structure at the sintered state (left) and the alternative heat exchanger structure (green state).

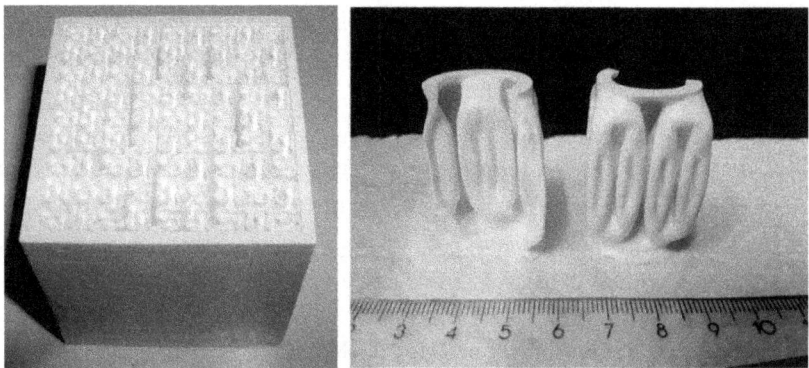

Figure 21. The FASS structures as ceramic component in the sintered (left; $35 \times 35 \times 35$ mm^3) or green state (right), additively manufactured via LCM technology.

5. Conclusion

The rapid development of AM technologies enables a radical paradigm shift in the construction of heat exchangers. In place of a layout limited to the use of planar or tubular starting materials, heat exchangers can now be optimized, reflecting their function and application in a particular environment. The investigations show the potential of the technologies concerning increasing heat exchanging surface and compactness as well as the designing of the fluidic systems. The AM of ceramics will pave the way to realize heat exchanger for operation temperatures highly above 1000°C.

The new approach for designing can also be used for bent structures. They provide more potential than the straight heat exchangers and open a wider field of possible technical applications. A possible, curved geometry is shown in **Figure 22**.

To increase the inner surface even more, rough surfaces, which induce beneficial turbulences, can be generated by modeling partial Brownian motion.

Figure 22. A computer-aided designed curved heat exchanger. Its geometry bases on FASS.

Acknowledgements

The authors would like to thank the German Federal Ministry of Education and Research (BMBF) for funding the project "**FunGeoS**" within the Framework Concept Zwanzig20— partnership for innovation" in the consortium AGENT-3D (fund number 03ZZ0208A) as well as European Union which funded parts of this work under the European Union's Horizon 2020 Research and Innovation Programme ("**CerAMfacturing**," Grant Agreement No 678503). Other parts of this chapter are based on results from **instaf**, a research project in the European network IraSME, which is carried out by partners in Austria and Germany. It is funded by the Federal Ministry for Economical Affairs and Energy (BMWi) on the basis of a decision of the German Bundestag.

Conflict of interest

There are no conflicts of interest and nothing else to declare.

Author details

Uwe Scheithauer[1]*, Richard Kordaß[2], Kevin Noack[3], Martin F. Eichenauer[3],
Mathias Hartmann[1], Johannes Abel[1], Gregor Ganzer[1] and Daniel Lordick[3]

*Address all correspondence to: uwe.scheithauer@ikts.fraunhofer.de

1 Fraunhofer Institute for Ceramic Technologies and Systems IKTS, Dresden, Germany

2 Fraunhofer Institute for Machine Tools and Forming Technology IWU, Dresden, Germany

3 TU Dresden, Team Geometric Modeling and Visualization, Dresden, Germany

References

[1] Tranter GmbH. Plattenwärmeübertrager. Available from: https://www.tranter.com/de/
products/plate-heat-exchangers [cited 13 October 2017]

[2] 3TRPD. Available from: https://www.3trpd.co.uk/ [cited 16 October 2017]

[3] Langefeld R, Veenker H, Schäff C, Balzer C. Additive Manufacturing—Next generation
(AMnx): Study. München. Available from: http://www.rolandberger.com/media/studies/
2016-04-11-rbsc-pub-Additive_Manufacturing-next_generation.html 11 April 2016

[4] Schnabel T, Oettel M, Müller B. Design for Additive Manufacturing: Guidelines and Case
Studies for Metal Applications. Ottawa: Fraunhofer-Institut für Werkzeugmaschinen und
Umformtechnik; 05/2017. Available from: http://canadamakes.ca/wp-content/uploads/
2017/05/2017-05-15_Industry-Canada_Design4AM_141283.pdf [cited 2017 9 October 2017]

[5] Choi SUS, Eastman JA. Enhancing thermal conductivity of fluids with nanoparticles.
ASME-Publications-Fed 1995;**231**:99-106 [cited 2017 Oct 11]

[6] Schatt W, Blumenauer H, editors. Werkstoffwissenschaft. 8., neu bearb. Aufl. Stuttgart: Dt.
Verl. für Grundstoffindustrie; 1996

[7] Mandelbrot B. The Fractal Geometry of Nature. New York: W.H. Freeman and Company;
1982

[8] Gebhardt A. Additive Fertigungsverfahren: Additive Manufacturing Und 3D-Drucken
für Prototyping-Tooling-Produktion. München: Carl Hanser Verlag GmbH Co KG; 2016

[9] Rombouts M, Kruth JP, Froyen L, Mercelis P. Fundamentals of selective laser melting of
alloyed steel powders. CIRP Annals—Manufacturing Technology. 2006;**55**(1):187-192.
DOI: 10.1016/S0007-8506(07)60395-3

[10] Bremen S, Buchbinder D, Meiners W, Wissenbach K. Selective Laser Melting—A
Manufacturing for Series Production. In: Fraunhofer Generativ, editor. DDMC—Direct
Digital Manufacturing Conference. 2012. p. 2012

[11] Viswanathan B, Aulice Scibioh M. Fuel Cells: Principles and Applications. Abingdon: Taylor & Francis Group; 2007

[12] Singhal SC, Kendall K. High-Temperature Solid Oxide Fuel Cells: Fundamentals, Design and Applications. Oxford: Elsevier; 2003

[13] Lee S, Kim H, Yoon KJ, Son J, Lee J, Kim B, Choi W, Hong J. The effect of fuel utilization on heat and mass transfer within solid oxide fuel cells examined by three-dimensional numerical simulations. International Journal of Heat and Mass Transfer. 2016;**97**:77-93

[14] Abel J, Scheithauer U, Klemm H, Moritz T, Michaelis A. Fused filament fabrication (FFF) of technical ceramics. Ceramic Applications. 2018;**6**:2-4

[15] Homa J. Rapid prototyping of high-performance ceramics opens new opportunities for the CIM industry. Powder Injection Moulding International. 2012;**6**:73-80

[16] Scheithauer U, Schwarzer E, Moritz T, Michaelis A. Additive manufacturing of ceramic heat exchanger: Opportunities and limits of the lithography-based ceramic manufacturing (LCM). Journal of Materials Engineering and Performance. 2018;**27**:14-10. DOI: 10.1007/s11665-017-2843-z

[17] Schwarzer E, Götz M, Markova D, Stafford D, Scheithauer U, Moritz T. Lithography-based ceramic manufacturing (LCM)—Viscosity and cleaning as two quality influencing steps in the process chain of printing green parts. Journal of the Electrochemical Society. 2017;**37**:5329-5338. DOI: 10.1016/j.jeurceramsoc.2017.05.046

Fluids

Thermal Performance of Shell and Tube Heat Exchanger Using PG/Water and Al$_2$O$_3$ Nanofluid

Jaafar Albadr

Additional information is available at the end of the chapter

http://dx.doi.org/10.5772/intechopen.80082

Abstract

This study investigates experimentally the thermal performance of propylene glycol/water with a concentration of (10/90) % and Al$_2$O$_3$/water nanofluid with a volume concentration of (0.1, 0.4, 0.8, 1.5, and 2.5) percentage under turbulent flow inside a horizontal shell and tube heat exchanger. The results indicate that the convective heat transfer coefficient of the nanofluid is higher than the base PG/water for the same inlet temperature and mass flow rates. The heat transfer of the nanofluid increases with the increase in mass flow rate as well as the Al$_2$O$_3$ nanofluid volume concentration. Results also indicate that the increase in the concentration of the particles causes an increase in the viscosity which leads to an increase in friction factor. The effect of Peclet number, Reynolds number, Nusselt number, and Stanton number has been investigated. Those dimensionless number values change with the change in the working fluid density, Prandtl number, and volume concentration of the suspended particles.

Keywords: heat exchanger, nanofluid, convection heat transfer

1. Introduction

With the rapid advancement in modern nanotechnology, nanofluids which are the working fluids with nanometer-sized particles (normally less than 100 nm) are used instead of micrometer size for dispersing in base fluids. The name was suggested for the first time by Choi [1]. The nanofluid can have significantly greater thermal conductivity with main disadvantage, which is increase in pressure drop. In this new age of energy awareness, reducing the size of the devices and increasing the efficiency are the main goal. Nanofluids have been demonstrated to be able to handle this role in some instances as a smart fluid. In electric power plants, the hot steam coming out from the turbine needs to be condensed, the unit that does this job called a condenser. Since its large size, the process of maintenance is difficult, is expensive, and needs a space for occupation. To minimize the size of the condenser, but still gets the same or better efficiency, we need to use a better working fluid than the water, and here comes the need of the nanofluid. The chapter is organized as follows: experimental setup is presented in Section 2. An explanation of the device used is introduced in Section 3. Experimental results are shown in Section 4. Finally, conclusion is outlined in Section 3.

2. Experimental setup

The project has been designed in [2] and used here to transfer heat from hot water in a heat exchanger to nanofluid and/or propylene glycol/water with a concentration of (10/90) % stored in a separate tank and make temperature calibrations for the same by employing two thermo-couples. Moreover, flow meters are installed in the pipes carrying nanofluid and PG/water to check the flowing rate. The complete system is very dynamic and easy to use. The mechanical structure design is shown in **Figure 1**.

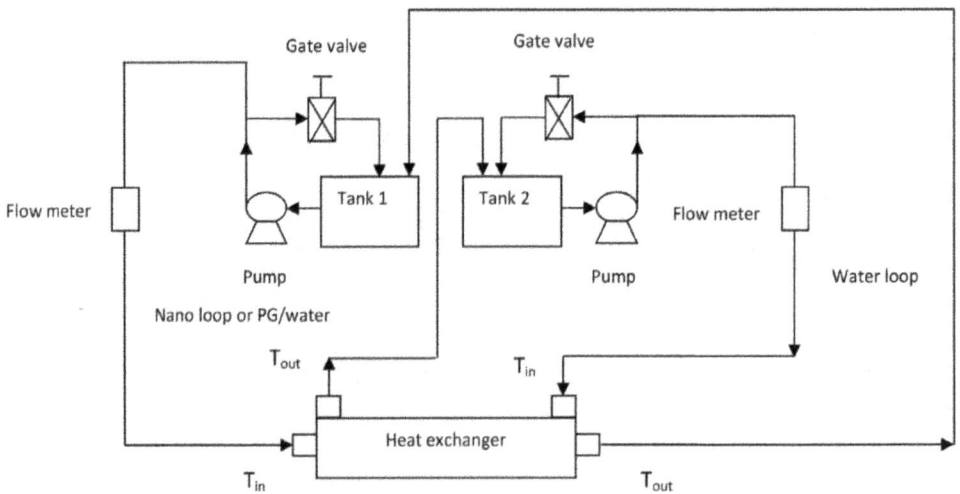

Figure 1. System diagram.

3. Basic working of the heat exchanger device

There are two loops in the system; the two flow loops carry heated nanofluid or PG/water and the other cooling water. Each flow loop contains a tank, a pump with a flow meter, and a bypass valve to choose the required flow rate for the study. The nanofluid tank (Tank 1) is filled with 4 liters, while the hot water tank (Tank 2) is filled with 12 liters. The 248-mm-long shell and tube heat exchanger is made of stainless steel type 316 L, with 37 tubes inside the shell. Each tube diameter is 2.4 mm, a wall thickness of 0.25 mm, and a heat transfer area of 0.05 m^2. Both inlet and outlet points of the liquid streams have a J-type thermocouples with removable bulbs for measuring the bulk temperatures. The experimental device was kept working for 15 minutes time period in each case of the chosen mass flow rates.

The system operates in a way that when the nanofluid or PG/water flows inside the tubes in the heat exchanger, the temperature reading will be calibrated that would come out to be normal room temperature. On the other hand, there is also another inlet valve that is connected to the tank where hot water is stored. This hot water also flows inside the shell of the heat exchanger device, and according to the law of thermodynamics, heat is transferred from the hot water to the nanofluid or PG/water that again gets back to the Tank 1 where it was stored initially through an outlet valve. The temperature reading is calibrated at both the nanofluid or PG/water inlet and exit points and to calibrate the readings from time to time according to the requirements. This reading is calibrated with the help of thermocouples. Thermocouple is a device that generates electric voltage at output proportional to the temperature readings.

4. Experimental results

To evaluate the accuracy of measurements, experimental system has been tested first with propylene glycol/water (10/90) % concentration before measuring the heat transfer characteristics of different volume concentrations of Al$_2$O$_3$/water. From the experimental system, the values that have been measured are the hot water temperatures of the inlet and outlet as well as the inlet of the PG/water and the different concentrations of nanofluid at different mass flow rates. The nanofluid presented equations are calculated by using the Pak and Cho [3] correlations, which are defined as follows:

$$\rho_{nf} = (1 - \varnothing)\rho_f + \varnothing\rho_p$$

The specific heat is calculated from Xuan and Roetzel [4] as the following:

$$(\rho Cp)_{nf} = (1 - \varnothing)(\rho Cp)_f + \varnothing(\rho Cp)_p.$$

Heat transfer rate can be defined:

$$Q = \dot{m} \cdot Cp \cdot \varDelta T$$

The logarithmic mean temperature difference:

$$\varDelta T_{lm} = \frac{(Twi - Tno) - (Two - Tni)}{Ln \frac{(Twi-Tno)}{(Two-Tni).}}$$

The overall heat transfer coefficient:

$$Q = U \cdot A_s \cdot \Delta T_{lm}$$

An alternative formula for calculating the thermal conductivity was introduced by Yu and Choi [5], which is expressed in the following form:

$$K_{nf} = Kf \frac{K\rho + 2Kf - 2\emptyset(Kf - K\rho)}{K\rho + 2Kf + \emptyset(Kf + K\rho)}$$

Thermal diffusivity is given by

$$\alpha_{nf} = \frac{k_{nf}}{\rho_{nf} \cdot c_p}$$

Drew and Passman [6] suggested the well-known equation of Einstein for calculating viscosity, which is applicable to spherical particles in volume fractions less than 5.0 vol.% and is defined as follows:

$$\mu_{nf} = (1 + 2.5\emptyset)\mu_w$$

The kinematic viscosity can be calculated from

$$\upsilon = \frac{\mu}{\rho}$$

Calculating Reynolds [7], Peclet and Prandtl [8] and Stanton numbers [9] is defined as follows:

$$Re = \frac{V.D}{\upsilon}$$

$$Pe = \frac{V.D}{\alpha_{nf}}$$

$$Pr = \frac{\upsilon_{nf}}{\alpha_{nf}}$$

$$St = \frac{Nu}{Re\ Pr}$$

Friction factors and Nusselt numbers for single-phase flow have been calculated from the Gnielinski equation [10] that is suitable for turbulent flow. The Gnielinski equation is defined as

$$f = [1.58(LnRe) - 3.82]^{-2}$$

Finding Nusselt number from the turbulent flow equation [10]:

$$Nu = \frac{(0.125f)(Re - 1000)Pr}{1 + 12.7(0.125f)^{0.5}\left(Pr^{2/3} - 1\right)}$$

Friction factor for each flow rate and nanoparticle concentration of the nanofluid can be found with the help of Duangthongsuk and Wongwises correlation [11] as Gnielinski equation [10] for single-phase flow cannot use for calculating friction factor as well as Nusselt number. Note that this equation is suitable for turbulent flow only and cannot be used for laminar flow:

$$f = 0.961 \cdot \text{Re}^{-0.375} \cdot \varnothing^{0.052}$$

Finding Nusselt number is calculated from Duangthongsuk and Wongwises correlation for turbulent flow [11]:

$$Nu = 0.074 \text{Re}_{nf}^{0.707} \cdot \text{Pr}_{nf}^{0.385} \cdot \varnothing^{0.074}$$

4.1. Enhancement ratio of thermal conductivity

From the calculated result, we can examine the enhancement ratio of thermal conductivity for PG/water as well as for the chosen concentrations of the Al$_2$O$_3$/water nanofluid.

The results show that thermal conductivity is a function of particle volume concentration; it increases with the increase in volume concentration. **Figure 2** shows the increase of thermal conductivity with respect to volume concentration of the nanoparticles. The maximum amount of thermal conductivity has been measured at 2.5% volume concentration with a value of 0.61629 W/m.k. This gives the maximum enhancement ratio of 1.0626. **Table 1** shows thermal conductivity enhancement ratio.

In compliance with thermal conductivity measurement, results show that the viscosity of the nanofluid increases with the increase in particle volume concentration and with decrease in temperature. The ability of nanofluid to absorb heat increases with the increase in volume

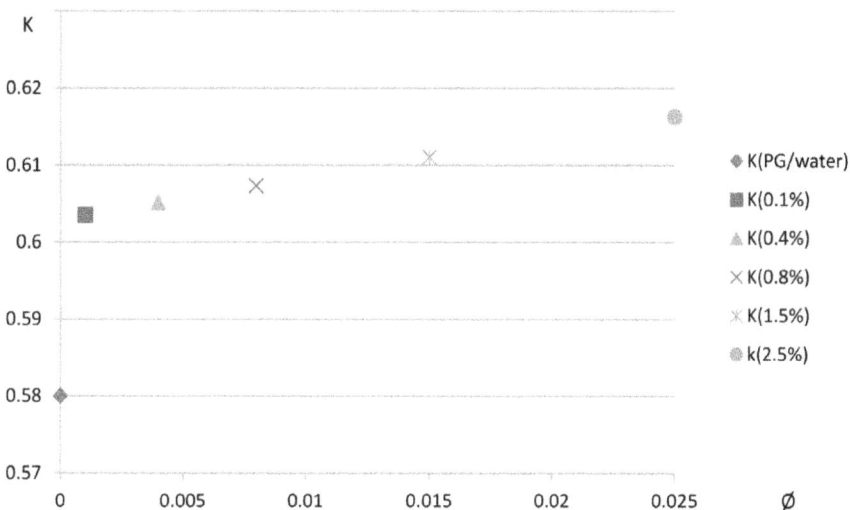

Figure 2. Thermal conductivity (K) versus volume concentration (Ø).

Thermal conductivity	Amount of K (W/m.k)	Enhancement ratio
K (PG/water)	0.58	—
K (0.1%)	0.60354	1.0405
K (0.4%)	0.60517	1.0434
K (0.8%)	0.60732	1.0471
K (1.5%)	0.61105	1.0535
K (2.5%)	0.61629	1.0626

Table 1. Enhancement ratio of thermal conductivity.

concentration of the nanoparticles that is why the density of the nanofluid remarkably increases with the increase in particle volume concentration of the nanofluid. This leads to increase the outlet temperature of the nanofluid, leading to raise the temperature difference of the nanofluid. For the specific heat, the results show that C_p decreases with the increase in volume concentration of the nanofluid. However, the amount of decrease in C_p is less than the amount of increase in density, which leads to reduction in thermal diffusivity with the increase in volume concentration of the nanoparticles in the nanofluid. The relationship between the calculated values of thermal diffusivity and kinematic viscosity can be seen in **Figure 3**.

Therefore, to apply the nanofluid for practical application, it is important to investigate their flow features in parallel with the thermal performance. In this study, suspended nanoparticles with volume concentrations of 0.1, 0.4, 0.8, 1.5, and 2.5 are utilized to calculate the friction factor for each volume concentration and for all the mass flow rates. **Figure 4** illustrates the values of the calculated friction factor with every measured value of Reynolds number. The results show that the friction factor increases dramatically with the increase in nanoparticle volume concentration for a given mass flow rate and decreases with increase in Reynolds number.

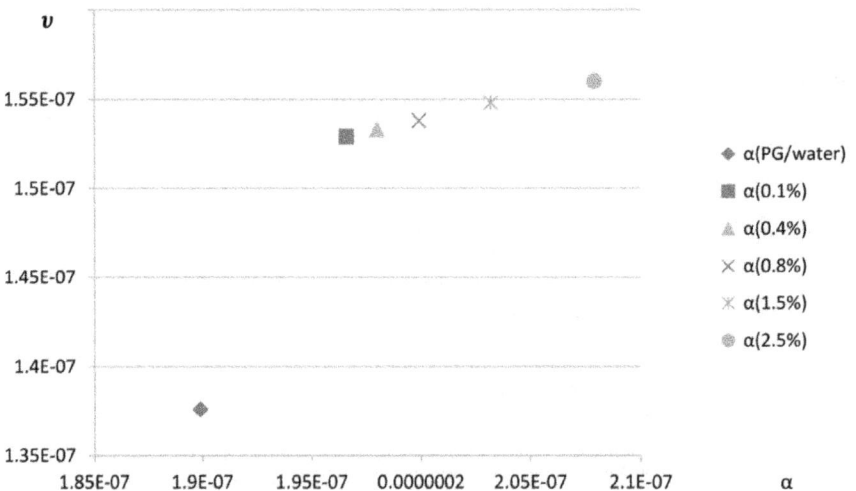

Figure 3. Thermal diffusivity (α) versus kinematic viscosity (υ).

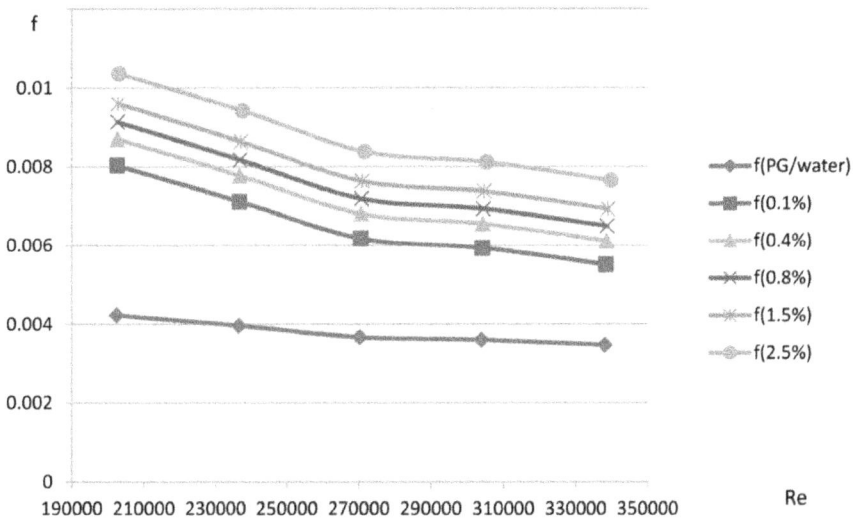

Figure 4. Friction factor (f) versus Reynolds number (re).

4.2. Enhancement ratio of heat rate

1. For ṁ = 30 L/min (**Table 2**)

Heat transfer rate	Amount of Q (W)	Enhancement ratio
Q (PG/water)	1514	—
Q (0.1%)	1519.5	1.0036
Q (0.4%)	1525	1.0006
Q (0.8%)	1540	1.0171
Q (1.5%)	1580	1.0435
Q (2.5%)	1624	1.0726

Table 2. Value of the enhancement ratio of heat rate at 30 L/min.

2. For ṁ = 35 L/min (**Table 3**)

Heat transfer rate	Amount of Q (W)	Enhancement ratio
Q (PG/water)	1843	—
Q (0.1%)	1848	1.0027
Q (0.4%)	1854	1.0059
Q (0.8%)	1875	1.0173
Q (1.5%)	1925	1.0444
Q (2.5%)	1950	1.058

Table 3. Value of the enhancement ratio of heat rate at 35 L/min.

3. For ṁ = 40 L/min (**Table 4**)

Heat transfer rate	Amount of Q (W)	Enhancement ratio
Q (PG/water)	2240	—
Q (0.1%)	2252	1.0053
Q (0.4%)	2264	1.0107
Q (0.8%)	2298	1.0258
Q (1.5%)	2320	1.0357
Q (2.5%)	2355	1.0513

Table 4. Value of the enhancement ratio of heat rate at 40 L/min.

4. For ṁ = 45 L/min (**Table 5**)

Heat transfer rate	Amount of Q (W)	Enhancement ratio
Q (PG/water)	2341	—
Q (0.1%)	2372	1.0132
Q (0.4%)	2390	1.0209
Q (0.8%)	2405	1.0273
Q (1.5%)	2465	1.0529
Q (2.5%)	2485	1.0615

Table 5. Value of the enhancement ratio of heat rate at 45 L/min.

5. For ṁ = 50 L/min (**Table 6**)

Heat transfer rate	Amount of Q (W)	Enhancement ratio
Q (PG/water)	3055	—
Q (0.1%)	3110	1.0180
Q (0.4%)	3155	1.0327
Q (0.8%)	3212	1.0513
Q (1.5%)	3222	1.0546
Q (2.5%)	3261	1.0674

Table 6. Value of the enhancement ratio of heat rate at 50 L/min.

The results show that the heat transfer rate of PG/water increases with the increase in mass flow rate till the amount of 45 L/min, and the heat rate starts to fall down at the mass flow rate of 50 L/min as the frictional forces became dominant. As for the nanofluid, the heat rate increases with the increase in mass flow rate and increases with the increase in volume concentration, a maximum enhancement ratio of 1.0674 is calculated at the maximum mass

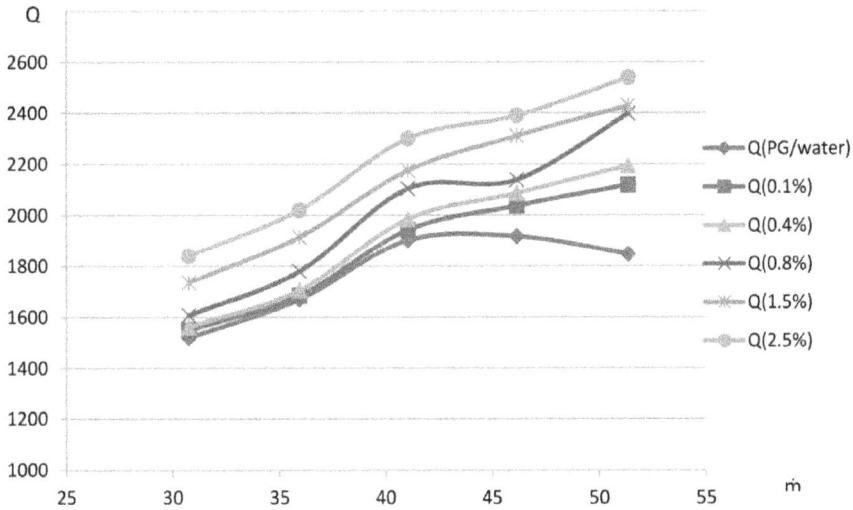

Figure 5. Heat rate (Q) versus mass flow rate (ṁ).

flow rate and at 2.5% volume concentration of nanofluid. **Figure 5** shows the pattern of the heat rate procedure at a different mass flow rates.

4.3. Enhancement ratio of the overall heat transfer coefficient

From the heat rate formula, the heat transfer coefficient is increasing with the increase of the heat rate (illustrated in **Figure 6**). The effect of heat transfer coefficient is directly proportional to the heat transfer rate.

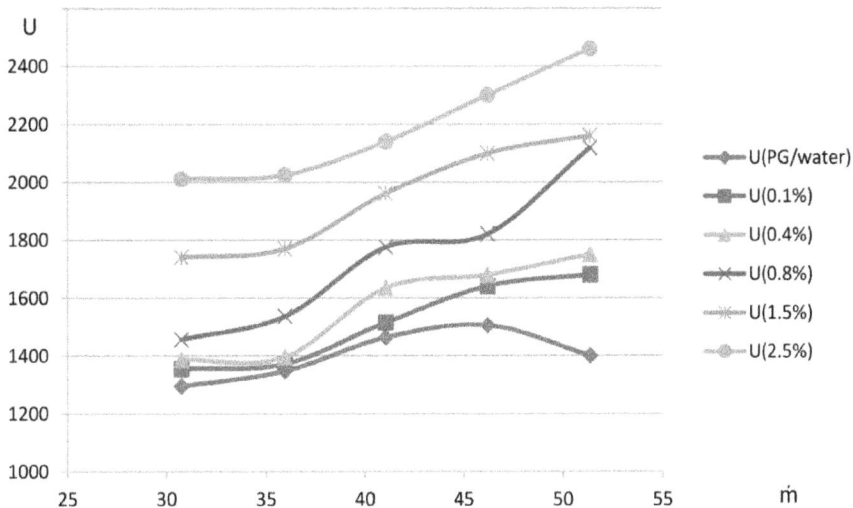

Figure 6. Overall heat transfer coefficient (U) versus mass flow rate (ṁ).

1. For ṁ = 30 L/min (**Table 7**)

Overall heat transfer coefficient	Amount of U (W/m²k)	Enhancement ratio
U (PG/water)	1295	—
U (0.1%)	1355	1.0463
U (0.4%)	1386	1.0702
U (0.8%)	1456	1.1243
U (1.5%)	1741	1.3444
U (2.5%)	2012	1.5536

Table 7. Value of the enhancement ratio of overall heat transfer coefficient at 30 L/min.

2. For ṁ = 35 L/min (**Table 8**)

Overall heat transfer coefficient	Amount of U (W/m²k)	Enhancement ratio
U (PG/water)	1347	—
U (0.1%)	1372	1.0185
U (0.4%)	1396	1.0363
U (0.8%)	1537	1.141
U (1.5%)	1770	1.314
U (2.5%)	2025	1.5033

Table 8. Value of the enhancement ratio of overall heat transfer coefficient at 35 L/min.

3. For ṁ = 40 L/min (**Table 9**)

Overall heat transfer coefficient	Amount of U (W/m²k)	Enhancement ratio
U (PG/water)	1463	—
U (0.1%)	1515	1.0355
U (0.4%)	1635	1.1175
U (0.8%)	1775	1.2132
U (1.5%)	1960	1.3397
U (2.5%)	2140	1.4627

Table 9. Value of the enhancement ratio of overall heat transfer coefficient at 40 L/min.

4. For ṁ = 45 L/min (**Table 10**)

Overall heat transfer coefficient	Amount of U (W/m²k)	Enhancement ratio
U (PG/water)	1505	—
U (0.1%)	1641	1.0903
U (0.4%)	1680	1.1162

Overall heat transfer coefficient	Amount of U (W/m²k)	Enhancement ratio
U (0.8%)	1820	1.2093
U (1.5%)	2099	1.3946
U (2.5%)	2300	1.5282

Table 10. Value of the enhancement ratio of overall heat transfer coefficient at 45 L/min.

5. For ṁ = 50 L/min (**Table 11**)

Overall heat transfer coefficient	Amount of U (W/m²k)	Enhancement ratio
U (PG/water)	1400	—
U (0.1%)	1680	1.2
U (0.4%)	1750	1.25
U (0.8%)	2118	1.5128
U (1.5%)	2160	1.5428
U (2.5%)	2460	1.7571

Table 11. Value of the enhancement ratio of overall heat transfer coefficient at 50 L/min.

Overall heat transfer coefficient and the heat rate shared the same thermal behavior. Initially, the overall heat transfer coefficient for propylene glycol/water increases as the increase in volumetric rate. However, the thermal energy starts reducing at a rate of 45 Ltr/min, as the frictional forces became dominant till it reaches the amount of 50 Ltr/min. The results indicate that when the mass flow rate as well as the nanoparticles volume concentration increases, the overall heat transfer coefficient of the nanofluid increases with a maximum value enhancement

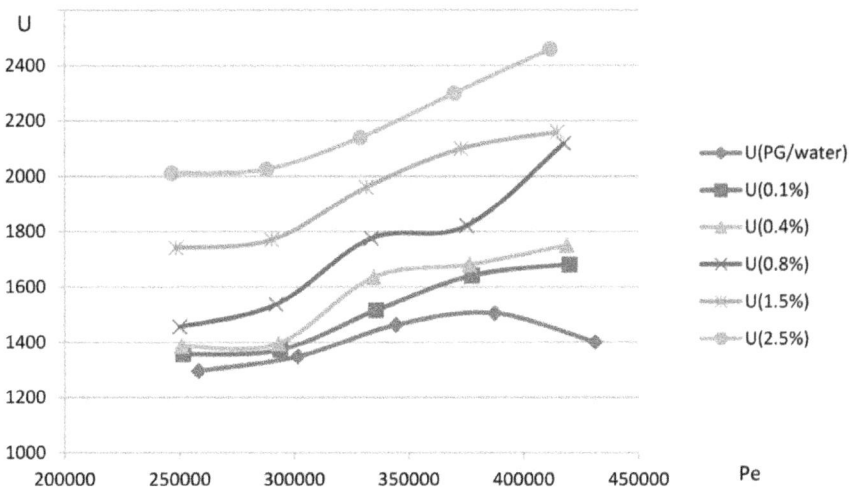

Figure 7. Overall heat transfer coefficient (U) versus Peclet number (Pe).

ratio of 1.7571 at a nanofluid with volume concentration of 2.5% and a volumetric rate of 50 L/min. The reason behind this increase in the thermal performance of the nanofluid is that the nanoparticles increase the thermal conductivity which leads to increase the heat transfer rate due to the chaotic movement of the nanoparticles. **Figures 6** and **7** show the process of the increase in heat transfer coefficient for a given mass flow rate and a given Peclet number.

4.4. Enhancement ratio of Nusselt number

1. For $\dot{m}$ = 30 Ltr/min (**Table 12**)

Nusselt number	Amount of Nu	Enhancement ratio
Nu (PG/water)	412.58	—
Nu (0.1%)	521.34	1.2636
Nu (0.4%)	577.16	1.3989
Nu (0.8%)	606.374	1.4697
Nu (1.5%)	632.437	1.5328
Nu (2.5%)	647.135	1.5685

Table 12. Value of the enhancement ratio of Nusselt number at 30 Ltr/min.

2. For $\dot{m}$ = 35 Ltr/min (**Table 13**)

Nusselt number	Amount of Nu	Enhancement ratio
Nu (PG/water)	458.03	—
Nu (0.1%)	582.08	1.2708
Nu (0.4%)	644.411	1.4069
Nu (0.8%)	677.47	1.479
Nu (1.5%)	706.315	1.542
Nu (2.5%)	723.1	1.5787

Table 13. Value of the enhancement ratio of Nusselt number at 35 Ltr/min.

3. For $\dot{m}$ = 40 Ltr/min (**Table 14**)

Nusselt number	Amount of Nu	Enhancement ratio
Nu (PG/water)	492.96	—
Nu (0.1%)	639.54	1.2973
Nu (0.4%)	708.06	1.4363

Nusselt number	Amount of Nu	Enhancement ratio
Nu (0.8%)	744.442	1.51
Nu (1.5%)	776.05	1.574
Nu (2.5%)	794.808	1.6123

Table 14. Value of the enhancement ratio of Nusselt number at 40 Ltr/min.

4. For ṁ = 45 Ltr/min (Table 15)

Nusselt number	Amount of Nu	Enhancement ratio
Nu (PG/water)	546.162	—
Nu (0.1%)	694.87	1.2722
Nu (0.4%)	769.312	1.4085
Nu (0.8%)	808.805	1.4808
Nu (1.5%)	843.155	1.5437
Nu (2.5%)	863.484	1.581

Table 15. Value of the enhancement ratio of Nusselt number at 45 Ltr/min.

5. For ṁ = 50 Ltr/min (Table 16)

Nusselt number	Amount of Nu	Enhancement ratio
Nu (PG/water)	590.18	—
Nu (0.1%)	749.19	1.2694
Nu (0.4%)	829.465	1.4054
Nu (0.8%)	872.06	1.4776
Nu (1.5%)	909.1	1.5403
Nu (2.5%)	931.09	1.5776

Table 16. Value of the enhancement ratio of Nusselt number at 50 Ltr/min.

The results show that the values of Nusselt number increase with the increase in the fluid velocity. The results also indicate that the nanofluid values of Nusselt number are higher than the base fluid, and these values are increasing with the increase in the nanoparticle volume concentration and Reynolds number. **Figures 8** and **9** represent the calculated values of Nusselt number and Stanton number for PG/water as well as for different concentrations of the Al$_2$O$_3$/water nanofluid in relation to the calculated values of Peclet and Reynolds numbers.

This behavior of the nanofluid is expected and has been recorded in [2], and it is due to the presence of the nanoparticles in the base fluid which increase the advantages like thermal conductivity and the molecular interchange as the molecules flows faster and the drawbacks

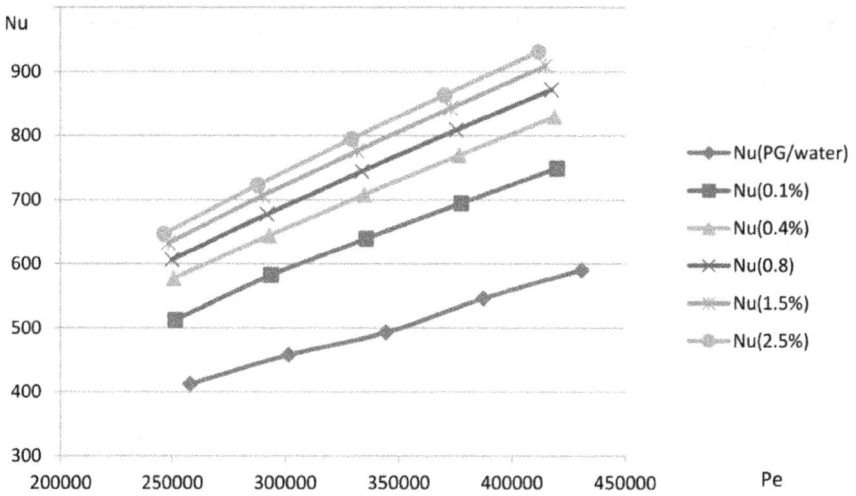

Figure 8. Nusselt number (nu) versus Peclet number (Pe).

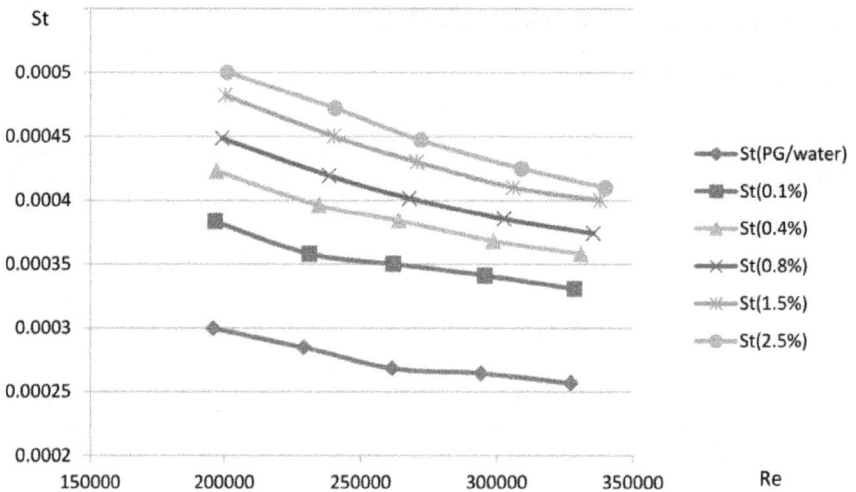

Figure 9. Stanton number (nu) versus Reynolds number (re).

mainly the viscosity at the same time. Heat transfer performance increases due to the increase in thermal conductivity, whereas the increase in viscosity leads to an increase in friction factor and increase in the boundary layer thickness. **Figure 10** illustrates the increase in the frictional forces versus the thermal energy enhancement (represented by Stanton number). It is worth mentioning that the temperature should be kept under the evaporation level to avoid the presence of vapor as the kinetic energy may exceed the binding energy causing the molecules to escape from the liquid.

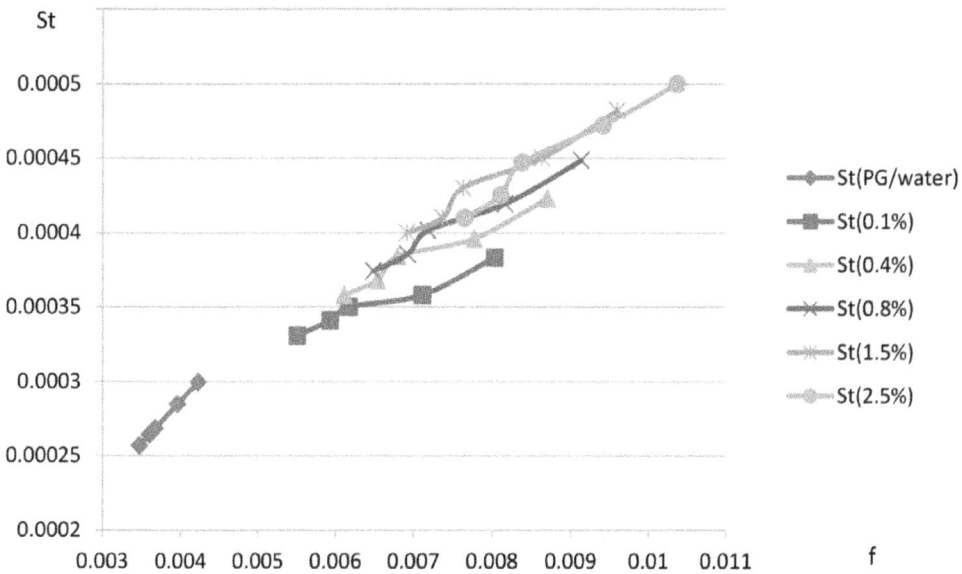

Figure 10. Stanton number (St) versus friction factor (f).

5. Conclusion

The convective heat transfer performance and flow characteristics of Al$_2$O$_3$ nanofluid flowing in a horizontal shell and tube heat exchanger have been experimentally investigated under turbulent flow. The effect of the Reynolds number and nanoparticle volume concentration on the flow behavior and heat transfer performance has been determined with different values of mass flow rates. Important conclusions have been achieved and are outlined as the following:

1. Thermal conductivity and viscosity increase due to the dispersion of the nanoparticles into the base liquid.

2. At a particle volume concentration of 2.5%, the use of Al$_2$O$_3$/water nanofluid gives significant higher heat transfer characteristics. For example, at the particle volume concentration of 2.5%, the greater enhancement ratio of heat transfer coefficient of the nanofluid at mass flow rate of 50 Ltr/min is 1.7571 which means that the amount of the heat transfer coefficient of the nanofluid is 57% greater than that of distilled water. As for Nusselt number, the maximum enhancement ratio at 50 Ltr/min is 1.5776. This means that Nusselt number of the nanofluid is 62.6% greater than that of distilled water.

3. Friction factor increases with the increase of a particle volume concentration. This is because of the increase in the viscosity of the nanofluid which means that the nanofluid incurs penalty in pressure drop which can lead to cavitation with higher volume concentrations.

Nomenclature

c_p	Specific heat (J/kg.k)
D	Tube diameter (m)
f	Friction factor
U	Overall heat transfer coefficient (W/m^2.k)
K	Thermal conductivity (W/m.k)
$\dot{m}$	Mass flow rate (Ltr/min)
Nu	Nusselt number
Pe	Peclet number
Pr	Prandtl number
Q	Heat transfer rate (W)
Re	Reynolds number
V	Mean velocity (m/s)
St	Stanton number

Greek symbols

Ø	Volume concentration (%)
ρ	Density (kg/m^3)
α	Thermal diffusivity (m^2/s)
μ	Viscosity (kg/m.s)
ΔT_{lm}	Logarithmic temperature difference (k)
υ	Kinematic viscosity (m^2/s)

Subscript

wi	Water inlet
wo	Water outlet
ni	Nanofluid inlet
no	Nanofluid outlet
in	Inlet
out	Outlet
n	Nanofluid

f   Base fluid

p   Nanoparticles

Author details

Jaafar Albadr

Address all correspondence to: jaafar.n.albadr@durham.ac.uk

Department of Engineering, University of Durham, Durham, UK

References

[1] Choi SUS. Enhancing thermal conductivity of fluids with nanoparticles. ASME FED. 1995;
     231:99

[2] Albadr J, Tayal S, Alasadi M. Heat transfer through heat exchanger using Al$_2$O$_3$ nanofluid
     at different concentrations. Case Studies in Thermal Engineering. 2013;**1**:38-44

[3] Pak BC, Cho YI. Hydrodynamic and heat transfer study of dispersed fluids with submi-
     cron metallic oxide particles. Experimental Heat Transfer. 1998;**11**:151-170

[4] Xuan Y, Roetzel W. Conceptions for heat transfer correlation of Nanofluids. International
     Journal of Heat and Mass Transfer. 2000;**43**:3701-3707

[5] Yu W, Choi SUS. The role of interfacial layers in the enhanced thermal conductivity of
     nanofluids: A renovated Maxwell model. Journal of Nanoparticles Researches. 2003;**5**:167

[6] Drew DA, Passman SL. Theory of Multi Component Fluids. Berlin: Springer; 1999

[7] Rott N. Note on the history of the Reynolds number. Annual Review of Fluid Mechanics.
     1990;**22**:1-11

[8] White FM. Viscous Fluid Flow. 3rd ed. New York: McGraw-Hill; 2006. pp. 69-91

[9] Webb RL, La C, Wisconsin. A critical evaluation of analytical solutions and Reynolds
     analogy equations for turbulent heat and mass transfer in smooth tubes. Wärme- und
     Stoffübertragung. 1971:197-204

[10] Gnielinski V. New equations for heat and mass transfer in turbulent pipe and channel
     flow. International Chemical Engineering. 1976;**16**:359-368

[11] Duangthongsuk W, Wongwises S. Heat transfer enhancement and pressure drop char-
     acteristics of TiO$_2$–water nanofluid in a double-tube counter flow heat exchanger. Inter-
     national Journal of Heat and Mass Transfer. 2009:2059

www.ingramcontent.com/pod-product-compliance
Lightning Source LLC
Chambersburg PA
CBHW081238190326
41458CB00016B/5825